MW00425798

PENGUIN BOOKS

# PAPER HEROES

Witold Rybczynski is the author of *Taming the Tiger, Home, The Most Beautiful House in the World,* a national bestseller, and, most recently, *Waiting for the Weekend.* He is a professor of architecture at McGill University and lives with his wife south of Montreal.

# PAPER HEROES

## Appropriate Technology: Panacea or Pipe Dream?

_____

*Witold Rybczynski*

PENGUIN BOOKS

PENGUIN BOOKS
Published by the Penguin Group
Viking Penguin, a division of Penguin Books USA Inc.,
375 Hudson Street, New York, New York 10014, U.S.A.
Penguin Books Ltd, 27 Wrights Lane,
London W8 5TZ, England
Penguin Books Australia Ltd, Ringwood,
Victoria, Australia
Penguin Books Canada Ltd, 2801 John Street,
Markham, Ontario, Canada L3R 1B4
Penguin Books (N.Z.) Ltd, 182–190 Wairau Road,
Auckland 10, New Zealand

Penguin Books Ltd, Registered Offices:
Harmondsworth, Middlesex, England

First published in the United States of America by
Anchor Books/Doubleday & Company, Inc. 1980
This edition with a new epilogue and new subtitle
published in Penguin Books 1991

1 3 5 7 9 10 8 6 4 2

Copyright © Witold Rybczynski, 1980, 1991
All rights reserved

ISBN 0 14 01.5375 6

CIP data available

Printed in the United States of America

# FOREWORD

The emphasis during what is sometimes referred to as the Machine Age has been most definitely on the devices and techniques which have distinguished this from other historical periods. The tendency for machines and technology to dominate society has become evident; what is less evident is how this relationship can be altered, if indeed it is not already too late to do so.

Appropriate Technology, or AT, is part lay religion, part protest movement, and part economic theory. If only because it combines such unlikely elements, it would be worth exploring, but there is more to it than that. In a period of growing disillusionment with technology and with the technological society, the Appropriate Technology movement attempts to grapple with a number of important issues: the relationship between technology and development, between ideology and industrialization, and, more fundamentally, between man and machine.

To date, the writing on these subjects has been by either proponents or practitioners of Appropriate Technology. In both cases there has been an understandable tendency to proselytize and pamphleteer—an instinctive glossing over of difficulties and highlighting of achievements. In both cases the problems associated with an evenhanded treatment of the subject have been compounded by the exigencies of the marketplace. The proponents have much to answer for here. In the rush to sell books to a youthful public they have frequently prevari-

cated, avoiding the truth when it was too painful and al-
tering it to suit their own, often commercial, aims. I have
more sympathy for the practitioners, who are, in many
cases, caught on the horns of a dilemma. How far should
they go in order to attract converts, how much should
they qualify an idea before it will lose all power to convince? All publications coming from a popular movement
are, to a certain extent, tracts, meant to rally support and
inspire action. Unfortunately, as a result of this tendency,
a distorted and one-sided picture has sometimes emerged.

"Skepticism is not sleep," the philosopher George San-
tayana once wrote, and though this book is frankly skeptical of a number of the pronouncements that the Appropriate Technology movement has made in recent years,
this does not necessarily imply that it is negative or reac-
tionary. It is a response, rather, to the excessive claims
and unsubstantiated promises of paper heroes, and to the
primitive nonsense that has been unscrupulously palmed
off as technology and science. It is, as a result, critical.
The purpose of this criticism, as regards adherents and
practitioners of Appropriate Technology, is definitely not
to convince them to abandon their endeavors (this would,
in any case, be presumptuous of me), but to attempt to
modify their views in the light of sometimes harsh reality.

I should make a confession at this point: I, too, write
as a practitioner. For a number of years my work with
the Minimum Cost Housing Group has, more often than
not, found its way into catalogs of appropriate technology, and hence I could be accused of biting the hand that
fed me. This may be ungrateful; it has not been, I hope,
dishonest.

I have discussed the topics examined in this book with
a number of friends and colleagues, both in Canada and

abroad. Among these I would like to record particular appreciation to Martin Pawley and to Vikram Bhatt, whose comments and criticisms have been most helpful.

My thanks also to Stewart Brand of the *CoEvolution Quarterly*, and Martin Spring of *Architectural Design*, who published my first articles on the subject of AT; to Bing Thom for his observations on China; to Robert Verrall of the National Film Board of Canada, for access to tapes of Schumacher interviews; to David Henry and Michael McGarry of the International Development Research Centre, for information on wind machines and sanitation technologies; to Professor Denton E. Morrison and Mario Kamenetzky for their papers on AT; to Alvaro Ortega; to Tom Lawand of the Brace Research Institute; to Willem Riedijk of the Dutch Appropriate Technology Centre and to Paul Osborn of TOOL for their views on technology and development; and to Dan Corsillo of the Architecture Photographic Laboratory of McGill University.

My appreciation to the staff of the McLennan Library, as well as to McGill University, in Montreal, which allowed me a leave of absence to complete this book.

At Doubleday my appreciation to Georgie Remer for her work in copy editing, and to my editor, Bill Strachan.

Finally, my thanks go to Shirley Hallam, my wife, for her suggestions and encouragement.

W.R.
*Montreal*
*September 1978–March 1979.*

# CONTENTS

# Chapter 1

# WHAT IS APPROPRIATE TECHNOLOGY?

"There is no human problem, indeed no field of human endeavor, that may not be approached and studied profitably through its history."

—JOHN LUKACS, *Historical Consciousness*

"Definitions," observed Samuel Johnson, "are tricks for pedants." Though, given the title of this chapter, the reader has a right to expect a definition of "Appropriate Technology," I am afraid he will be disappointed. A dozen or so different books have given a dozen or so varying definitions.[1] My purpose is not to add yet another, but to

[1] Here is a baker's dozen; most of them, though not all, are referred to in the text or in footnotes: E. F. Schumacher, *Small Is Beautiful: Economics As If People Mattered* (London, 1973); Bruce McCallum, *Environmentally Appropriate Technology: Developing Technologies for a Conserver Society in Canada* (Ottawa, 1973); David Dickson, *Alternative Technology and the Politics of Technical Change* (London, 1974); Ken Darrow and Rick Pam, *Appropriate Technology Sourcebook* (Stanford, Calif., 1976); Peter Harper et al., eds., *Radical Technology* (New York, 1976); Canadian Hunger Foundation & Brace Research Institute, eds., *A Handbook on Appropriate Technology* (Ottawa, 1976); N. Jéquier, ed., *Appropriate Technology: Problems and Promises* (Paris, 1976); Ignacy Sachs et al., *Techniques Douces, Habitat et Société* (Paris, 1976); Richard S. Eckaus, *Appropriate Technology for Developing Countries* (Washington, D.C., 1977); R. J. Congdon, ed., *Introduction to Appropriate Technology: Toward a*

examine the reason for the diversity of the preceding ones. The stubborn reader is nevertheless encouraged to consult a dictionary; he will not be enlightened. "Appropriate" means something "attached as a peculiar attribute or quality" and, more generally, "specially suitable or proper." It is fair to ask how a technology can be "specially suitable," or rather, since every technology is suitable for something, how does "appropriate technology" differ from simply "technology"?

George Orwell pointed out some years ago the debasement of meaning in language. The "double-speak" which he fantasized in a novel has become very much a part of the modern world. Ministries of war are called "ministries of defense," spy organizations are referred to as "the intelligence community," the police are described as "security forces." Nor has this convolution been restricted to the political sphere. Ecology, which actually means the study of the relation of living organisms to their surroundings, is used incorrectly to refer to the environment as a whole, and, by a feat of propagandizing zeal, "ecological" has come to mean "beneficial to man's environment." And this imbuing of words with hidden, and sometimes not so hidden, values is often done by scientists who should know better.

There is some of this propagandizing tendency in the use of "appropriate" or "alternative" technology, terms which are loaded with veiled meanings, the exploration of which is the purpose of this chapter. Through a happy accident of syntax, "AT" has come to refer to both alterna-

---

*Simpler Life-style* (Emmaus, Pa., 1977); Amory B. Lovins, *Soft Energy Paths: Toward a Durable Peace* (Cambridge, Mass., 1977); J. Baldwin and Stewart Brand, eds., *Soft-Tech* (Sausalito, Calif., 1978); Lane De Moll and Gigi Coe, eds., *Stepping Stones: Appropriate Technology and Beyond* (New York, 1978).

tive technology and appropriate technology. Much as I would prefer to use a more accurate terminology, it is "AT" which has come to the fore and which is widely used, and so it is "AT" to which I will refer, implying a specific movement which exists quite apart from the meaning, or lack thereof, in the original etymology.

The idea of "appropriate technology" could be said to have emerged publicly at a meeting which took place at St. Cross College, Oxford University, England, in 1968. In spite of the setting, and the convention's rather cumbersome title—"Conference on the Further Development in the United Kingdom of Appropriate Technologies for, and Their Communication to, Developing Countries"— this was not a meeting of academics. Rather, it brought together captains of British industry, representatives of such international organizations as the United Nations Economic Commission for Africa, the International Labor Organization, and the powerful Organization for Economic Cooperation and Development, as well as officials of the British Ministry of Overseas Development. There were more than one hundred participants.

This conference had been organized by a small private body, half charity, half pressure group, called the Intermediate Technology Development Group. Its aim was to alert British industry to the technological needs of the less developed countries, those states in Asia and Africa which had recently acquired independence and were beginning the difficult transition from primitive agrarian to mechanized societies. In particular it was hoped that the industrialists could be made aware of what had been identified as a need for simpler, more labor-intensive "intermediate" technologies.

It is difficult to tell, reading between the lines of the

polite speeches, what impact the notion of intermediate technology had. In any case, this will be dealt with later; for the moment, it is interesting to note that one of the participants ventured to remark that "in fact 'appropriate' might be a better word than 'intermediate.' In this context, we had to dispel fears that we're thinking in terms of second best."

It appears that considerations of public relations won out over semantic accuracy, for five years later the Intermediate Technology Development Group announced that it was changing the name of its newsletter, the *IT Bulletin*, to *Appropriate Technology Journal*. The announcement of this change took place during another meeting, this one entitled "Appropriate Technology." Unlike its predecessor, this conference was dominated by academics, and consequently a certain amount of the discussion centered on problems of definition, criteria, and the like. An Indian visitor, M. K. Garg, stated, "So many people with diverse orientations and motivations have gathered under this banner [AT] that one sometimes wonders whether it is not a mere slogan raised by those who find themselves left out of the mainstream of the development process." Interestingly, these words of caution came from one of a handful of non-European participants. If his rather caustic description of appropriate technologists as outsiders fitted most of the participants, it certainly did not apply to the founder-director of the Intermediate Technology Development Group, E. F. Schumacher. This obscure economist, a recently retired British civil servant, was an unlikely candidate for the role of charismatic leader of what was to become known as the "Appropriate Technology movement."

## E. F. SCHUMACHER

Ernst Friedrich (Fritz) Schumacher was born in 1911 in the famous university town of Bonn where his father was professor of political economy.[2] The young Schumacher grew up in what was a turbulent period for Germany—the disastrous World War I, the ignominy of defeat, rampant inflation, and, finally, in the early twenties, the rise of National Socialism. After finishing his studies at Bonn and Berlin, Schumacher went to Oxford as one of the first group of German Rhodes Scholars since the war. He subsequently went to Columbia University in New York where his father had been the first Kaiser Wilhelm Professor thirty years before; following quite literally in his father's footsteps, Schumacher also became an economist. In 1934 he returned to Germany to begin a career in import-export; that was the year that the aged President Paul von Hindenburg died and was replaced by Chancellor Adolf Hitler. The Nazification of German life began: politics, education, law, and business were affected. By the end of 1938 the Third Reich had swelled to include the Rhineland, Austria, and, in effect, Czechoslovakia, but Schumacher was not there to witness the golden age; like many German intellectuals he had decided to emigrate, though unlike most (Albert Einstein, Walter Gropius) he did not choose the United States. In 1937, newly married and at the age of twenty-six, he arrived in England.

[2] Hermann A. Schumacher had a long and particularly distinguished career. He founded the commercial university of Cologne, and was a professor of political economics at both the universities of Bonn and Berlin. He was a prodigious traveler and visited China, Japan, and Korea as early as 1896; later trips took him to Java, Sumatra, and Malaya. His reputation may be judged by the fact that he was listed in the *International Who's Who* for over a decade.

The reader is entitled to ask at this point why this brief excursion into personal history in a book that intends to deal with ideas and not personalities? There are two reasons. First, E. F. Schumacher was undoubtedly the motive force behind the AT movement. It is not an exaggeration to say that without him there would have been no AT. At the same time, some of the contradictions of this movement were those of Schumacher himself, and a glance at the man, however cursory (and this is no biography), does shed light on his ideas. Second, given his pivotal role in the development of this public idea, it should be understood that Schumacher was neither a crackpot media-guru nor a typical word-of-mouth celebrity. If his views were contentious, they are nevertheless worth discussing, for they were the views of a remarkable and complex person.

When Schumacher arrived in England he continued his career in business by starting a company that marketed electric delivery vehicles.[3] This enterprise was short-lived, however, as World War II broke out two years later. One of the wartime measures that came into effect shortly thereafter was the internment of German nationals. Schumacher's internment lasted three years; he spent it working as a farm laborer in Northamptonshire. One can only speculate what effect blue-collar rural life had on the young German businessman, but since Schumacher alluded to it several times in his later life ("this was my main university"), it must have been not inconsiderable. Following his internment, Schumacher worked as a jour-

[3] It is probable that the abrupt move from Germany to England had an important influence on Schumacher. In an interview recorded by the National Film Board of Canada three and a half months before his death in 1977, he said, "Economists talk about mobility as a wonderful thing; well, maybe it means rootlessness. Maybe it means you don't belong anywhere. I have myself a very great need for roots."

nalist for the London *Times* and briefly at Oxford University as a member of the staff that prepared the Beveridge Plan, a report which outlined a new social security system that was to become the basis for the welfare state legislation introduced by the Labour government in 1945, and finally as a member of the United States Strategic Bombing Survey. In 1946 he joined the British Control Commission in Germany, where he spent four years as an economic adviser.

The main part of Schumacher's working life was spent as a technocrat and civil servant.[4] From 1950 to 1970 he worked for the British National Coal Board, first as an economic adviser and later as director of statistics. His career seems to have been uneventful. Although coal had been the backbone of the Industrial Revolution, peak production was passed as early as 1914, and with the advent of petroleum a steady, inevitable decline set in. By the time it was nationalized in 1947, the British coal industry was shrinking rapidly. During the sixties nearly half of the coal industry was abandoned as uneconomical. The warnings of people like Schumacher that "the National Coal Board has one overriding task and responsibility, being the trustees of the nation's coal reserve; to be able to supply plenty of coal when the world-wide scramble for oil comes" were ignored.[5]

[4] I do not use the term "technocrat" in a pejorative sense, as is now the fashion. A technocrat is a "technical expert who establishes principles on which a particular social order is based." This precisely describes the role of a civil-servant economist.

[5] *Prospect for Coal* (London, 1969). But as early as 1958, Schumacher had warned that European dependency on the Middle East for oil could have serious political repercussions. He was (justifiably) proud of his foresight and showed this report to a number of interviewers (*New Scientist*, September 12, 1974; *Building Design*, July 12, 1974).

If Schumacher's career with the National Coal Board contributed little except a growing pessimism about the foresight of modern industry, it did involve him, as the senior economist of a large state-owned industry, in serving as an adviser to overseas governments. One of the first opportunities came in 1955 when he spent three months in Burma as a United Nations economic adviser to the local government.[6] His experiences in this and other Asian countries prompted his concern, as an economist, for the plight of the poor. He was later fond of saying, "I am interested in the poor; I've always found that the rich can look after themselves, they don't need me." It was in 1963, following a visit to India, that Schumacher coined the term "intermediate technology," and three years later he founded the Intermediate Technology Development Group.

## INTERMEDIATE TECHNOLOGY

The question has sometimes been asked, "Intermediate between what?" Schumacher's original definition was quite clear. He described the need of the poor in less developed societies for a technology that was "more productive than the indigenous technology but immensely cheaper than the sophisticated, highly capital-intensive technology of modern industry."[7] In order to be successful, according to Schumacher, this technology, which he called "intermediate," should have four characteristics. First, it should create employment in the rural areas to re-

[6] The Burmese experience is important. Burma is one of the very few countries that has actually resisted economic progress, a unique example of voluntary poverty in the so-called Third World.

[7] Small Is Beautiful: A Study of Economics As If People Mattered (London, 1973).

duce the urban migration that characterized most developing countries. Second, this technology should be
labor-intensive rather than capital-intensive; that is, it
should create jobs with the minimum of investment. This
was aimed at reducing the high unemployment found in
many less developed countries. Third, methods employed
should be "simple" so as to reduce demand for skilled
labor and management. Fourth, production should be, as
much as possible, for local use.

The theory of intermediate technology was proposed by
Schumacher at a particularly critical moment. Notwithstanding the fact that all poor countries were referred to
as "developing," the unpleasant and depressing fact was—
and is—that many of these countries were not developing
at all. The United Nations Development Decade had been
hopefully announced in 1964. The stated aim was for all
the poor countries to increase their per capita income by
50 per cent (5 per cent per year). This would have still
left the poor countries far behind the industrial nations,
but it was considered a bare minimum of the change necessary. By the early seventies, statistics indicated that although the average growth had indeed slightly exceeded
the target figure set down by the United Nations, fully two
thirds of the less developed countries had experienced a
growth rate that was significantly lower than 4 per cent,
and in some cases they had even suffered a decline.[8] The
optimism of plans such as the Development Decade was
not completely without foundation. The Marshall Plan

[8] For the period 1965–74 the World Bank indicated that in
South America only Brazil had an annual gross national product
(GNP) per capita growth of over 5 per cent; in Africa only oil-
rich Nigeria and Libya exceeded the figure; and virtually none of
the countries in Asia, which includes the giants India and China,
achieved the target.

of 1948–51 had had an extravagant effect on postwar Europe. Countries which had been bombed "back to the Stone Age," to use a phrase from a later war, revived and prospered as they had never done before. Surely, it was felt, the same could be done for the poor countries, many of which had become independent in the fifties and early sixties.

One such effort was the so-called green revolution.[9] This term referred to the development of various high-yield crops for the tropics, chiefly rice and wheat. These were introduced into a number of less developed countries, together with modern agricultural techniques such as controlled irrigation, fertilization, and weeding. The initial results of the green revolution, greatly publicized, were almost too good to be true: wheat production in certain countries doubled, rice harvests tripled. The first countries to feel the benefits were Mexico, India, Pakistan, and the Philippines—a not inconsiderable experiment. The green revolution was the prototype for solving the problems of the developing countries with modern technology, for, though the high-yield rice and wheat were developed in the Philippines and Mexico, the green revolution was very much a Western approach, involving both the Rockefeller and Ford foundations.

It became apparent, however, that the green revolution was not a panacea; as with all development, new problems had to be faced. Particularly in the early stages, the beneficiaries of the innovative agriculture were the

[9] The "green revolution" was named by a U. S. Agency for International Development (AID) official, presumably in optimistic and rather simpleminded imitation of the Red (Russian) Revolution. Ironically, there are indications that in some countries, such as India, the green revolution has actually accelerated the desire for rapid social reform.

larger and richer farmers who could afford to invest in the required seeds, irrigation, and fertilizers. The optimistic production figures of the first phase were often the result of planting the best land first; later production tended to drop. In spite of the fact that enormous progress was made in countries like India, where food production had previously been stagnating, the green revolution was not equally successful in all countries. In many cases, after the initial burst of activity, government support for the agricultural programs dwindled; the research that was necessary to sustain a modern agriculture was not continued. It was easy to criticize the pretentiously named green revolution, and many did so, particularly those from advanced rather than less developed countries.[10] Although some of these criticisms were no more than scholarly carping, others had validity, particularly in those cases where improved agricultural technology was adopted without significant parallel agrarian reform.

The result of partial failures such as the Development Decade or of perceived failures such as the green revolution created a climate of pessimism as regards the ability of technology to solve the pressing problems of the poor countries. Of course, the extent of this pessimism should not be exaggerated. The Soviet Union still gave a hydroelectric dam to Egypt, Canada built nuclear reactors in India, and China constructed a railway in Mozambique. Even as advanced technology was criticized, it was apparent that it remained the only way to progress, and for most less developed countries, the only desired way.

Nevertheless, when Schumacher came on the scene

[10] For instance, N. Wade, "Green Revolution: A Just Technology Often Unjust in Use" *Science,* Nos. 185 and 186, December 1974, and Wolf Ladejinsky, "Ironies of India's Green Revolution" *Foreign Affairs,* July, 1970.

with the "intermediate technology" proposal, there were
many who were willing to listen, first, because modern
technology could obviously not solve all the problems, especially in the rural areas, as the green revolution had
shown. Second, as the Development Decade had so graph-
ically illustrated, it was unlikely that most of the less
developed countries would be able to modernize in a
hurry, the reconstruction of Europe notwithstanding.
Thus, even when industrialization was considered to be
the ultimate goal, intermediate technology was seen as a
desirable technique for bridging the gap, a stepping stone
to modernization.

By the early seventies the intermediate technology approach was beginning to be accepted as an adjunct to traditional theories of development. Though a certain number of small groups which lobbied for more emphasis on
rural development and small-scale technology had sprung
up, there was no general "movement" in this direction.
Perhaps because the ideological base of intermediate technology was narrow,[11] perhaps because it was aimed solely
at the less developed countries, or perhaps because it really was seen as second best, the impact of intermediate
technology, particularly on the public, was very limited.

## SMALL IS BEAUTIFUL

The publication, in 1973, of *Small Is Beautiful: A
Study of Economics As If People Mattered* propelled
Schumacher into the limelight. This collection of Schumacher's lectures and essays, "carpentered together into a
single book" as the London *Times* reviewer put it, be-

[11] For this reason, intermediate technology was never popular
with the American youth movement or with the left; it was both
unglamorous and too compromising.

came an international bestseller and was translated into fifteen languages. It bears a remarkable similarity to another bestseller, Betty Friedan's *The Feminine Mystique*. Both books are ponderous (Schumacher's, less scholarly, makes no pretense at continuity at all), yet both have served as cultural milestones. *Small Is Beautiful* was quoted by figures as disparate as the heir to the British throne, Prince Charles, and an heir of another sort, Governor Jerry Brown of California.[12] If Schumacher's name was hardly to become a household word, "small is beautiful," at least, would take its place in the true popular literature of the Modern Age—the T-shirt slogan.

The success of *Small Is Beautiful* can probably be attributed to its most serious limitation: it did not attempt a reasoned argument but appealed directly to the emotions. Since it was a collection of essays on a variety of subjects, it gave the impression of covering a lot of ground, and even though some of the statements were contradictory (which is inevitable in a book, by a single author, that contains material spanning more than a decade), it offered simple and understandable solutions.

What was this book really about?

[12] It is likely that many of the public figures who have endorsed the "small is beautiful" outlook have not actually read the book. The following exchange reportedly took place at a public meeting attended by Governor Brown: "About ten minutes into the question period, a young man in the front rows asked the governor what he thought about the last two chapters of Schumacher's book —the ones where he called for gradual conversion to a system of decentralized socialism, with the workers and the communities controlling the factories. 'Could you repeat that?' Brown asked, although it was entirely audible. 'I'm not sure I understand.' The student repeated the question. 'Well I'm not sure I know what you're referring to. Those last chapters are kind of vague.' An interesting response considering the fact that the last two chapters of *Small Is Beautiful* are *the* most specific in the book" (Joe Klein, "The Prince of the West," *Rolling Stone*, No. 217, July 15, 1976).

*Small Is Beautiful* was first and foremost a diatribe against modernization: the bureaucratization and rationalization of all aspects of modern life, the depletion of nonrenewable resources, environmental decay, even the Bomb. This in itself was hardly news; there had been a series of books that had done this more comprehensively, and sometimes more convincingly, beginning with Jacques Ellul's *The Technological Society* (New York, 1964). Schumacher took the moral and material decay of the modern world pretty much for granted (as did most of his readers, one suspects), and the appeal of his message rested on two other attributes.

Most critiques of modernization had been confined to Western technological society. Schumacher pointed out, perhaps for the first time, the link that existed between the discontents in the West and those in the less developed countries. The latter discontents can be characterized as movements toward countermodernization, most frequently evidenced by traditionalism (the Middle East) or nationalism (Africa). Demodernization and countermodernization spring from different sources; vastly oversimplified, the former could be said to be of the left, the latter of the right. Though one is a reaction against modernization and the other is fear of an alien imposition, they are both engaged in a rebellion against modernization and its effects.

The second appeal of Schumacher's book is the fact that, unlike Ellul, a French sociologist, he did offer a solution to the malaise of modernization, and, in spite of his disclaimers, it appeared to be a technological solution. A conundrum emerged from this book and became more evident in Schumacher's later writing. All of Schumacher's training and experience had made him a technocrat; perhaps a particularly sensitive technocrat, but a techno-

crat nevertheless. Schumacher was an economist and he
tended to see the solutions to problems in economic
terms. His remedies for the self-destructive tendencies of
modernization consisted in looking for answers within the
technocracy itself; changing the kind of decision that technocrats make; choosing, for instance, smaller instead of
larger industries. For these reasons, *Small Is Beautiful*
elicited a strong positive response from state institutions.

It is tempting to compare Ernst Friedrich Schumacher
with another German economist, Karl Heinrich Marx
(1818–83). The superficial similarities, though coinci-
dental, are curious. Both came from upper-middle-class
backgrounds; they were born within seventy miles of each
other. They attended the same two German universities,
Bonn and Berlin. They both had large families (Marx six
children, Schumacher eight), both left Germany and spent
the most productive part of their careers in England
(where both are buried), and both wrote books on political economics. Schumacher, however, was no Marxist.
It is unlikely that he was even a socialist; his endorsement of socialism in *Small Is Beautiful* was tentative indeed, since he recognized that most forms of socialism
share the same attitude to technology as do most capitalist societies.[13] Nevertheless, one has to conclude that
Schumacher did largely share Marx's doctrine of "economic determinism." I have already pointed out that
*Small Is Beautiful* was a collection of lectures and essays
which made for uneven and sometimes inconsistent read-

[13] Questioned about his combination of conservative and revolutionary views, Schumacher responded, "That may simply mean
that the traditional categories of conservative, or conformist, or
revolutionary don't apply anymore . . . and the cake is being cut
in different directions now" (National Film Board of Canada interview, 1977).

ing. It is difficult to make conclusive statements about the views it contains, since often opposing sides of an argument are supported in two different chapters. However, the impression of the book as a whole, and certainly its popular impact, was economically and technologically deterministic; that is, it assumed that the modern world was shaped less by politics, geography, and national culture than by economics and technology.

Schumacher, a complicated man, was also a complicated economist. He grew up in post-World War I Germany in the same region as did the German youth movement which originated in 1896 but continued in various forms until 1938, encompassing the sons, and later the daughters, of the middle class. This was a unique occurrence, unparalleled in the rest of Europe and not to be repeated until the "hippies" of the 1960s. The *Jugendkultur* was apolitical and pacifist and included such things as whole-food diets, exercise, and vegetarianism, all cast in a romantic criticism of bourgeois society. If the reader sees a similarity between this movement and the Californization of present-day youth culture, he is not mistaken. Eastern religion, mysticism, folk singing, communes, and free schools were all part of the German youth movement. A favorite German author of the 1920s, Herman Hesse, underwent a revival in the 1960s in the United States and Canada, if not exactly for the same reasons, at least with the same age group.[14] It would be surprising if

[14] The American youth culture has become the focus of a certain amount of attention as a result of Charles A. Reich's *The Greening of America* (New York, 1970) and Theodore Roszak's *The Making of a Counter Culture* (New York, 1969). It is worthwhile to contrast these ingratiating and opportunistic histories with a look at the earlier German youth culture by the historian Walter Laqueur, who had, let it be granted, the benefit of hindsight: "What did Chinese philosophy and Indian mysticism mean to the

Schumacher were not also influenced by the ideals of the
*Jugendkultur*. Much of what later went toward making
Schumacher a cult figure—his organic gardening, his paci-
fism, even his Buddhism—were by no means a pose.
Through a series of historical accidents Schumacher
bridged the gap of forty years that separated the American
hippies from the German *Wandervögel*. The moral, sometimes mystical, side of Schumacher attracted the American *Jugendkultur* to AT, but it did something else as
well. Because it was genuine, not an affectation, it created
a very real tension in Schumacher between the moralist
and the economist.[15]

Schumacher's religious development was greatly affected, by his own admission, by the visit he paid to
Burma, which followed his earlier interest in mysticism
and psychic research. (Rudolf Steiner, the inventor of "an-
throposophy," was a big influence on German youth of the
twenties and thirties, and Schumacher frequently lectured
at a Steiner school in England.) In *Small Is Beautiful*

---

generation of 1918–19? Strictly speaking, nothing at all; but they
liked the parables; the fact that these were not at all applicable at
time of political and social crisis to people with an entirely
different cultural tradition in the heart of Europe, was a secondary
consideration" (*Young Germany*, London, 1962).

[15] This tension, like others in Schumacher's life, was oddly
self-inflicted. He went not to the United States, "the land of opportunity," but to England, where in one sense he would always
remain a foreigner (his honorary doctorate was from a German,
not an English, university). Later, living in England, he chose to
become a Roman Catholic—in a Protestant milieu the religion of
the outsider (of G. K. Chesterton, Hillaire Belloc, Graham
Greene). At the same time, like many Germans of his generation,
he was enough of an Anglophile to be described by an American
interviewer as "tall, elegant and tailored by Savile Row . . . he
still looks like one of those benign silver-haired aristocrats who
offer Scotch to the discriminating in glossy magazine ads" (New
York *Times*, October 26, 1975).

Schumacher introduced the idea of Buddhist economics, which consisted basically of applying restraint and self-discipline to economic decision making.[16] This was an attempt to try and resolve the dilemma between the moralist and the economist. The moralist wanted to change man, the economist wanted to change the social system. This dilemma was never satisfactorily resolved. In his last book, *A Guide for the Perplexed* (London, 1977),[17] there are indications that Schumacher-the-economist had deferred to Schumacher-the-moralist: "Even if all the 'new' problems were solved by technological fixes, the state of futility, disorder, and corruption would remain. It existed before the present crisis became acute, and it will not go away by itself." This despairing statement, however, was made four years later; at the time that *Small Is Beautiful* appeared, it was still the economist who predominated. The ills of society were caused by technology, he wrote; to correct these ills it was necessary to choose a different kind of technology, a "technology with a human face."

There were very few examples in *Small Is Beautiful* of what this technology might be, or indeed if it *could* be—the bulk of the book dealt with what *should* be. These human-faced machines were to be "cheap enough so that they would be accessible to virtually everyone; suitable for small-scale application; and compatible with man's need for creativity." If we would only develop machines

[16] The term "Buddhist economics," to which an entire chapter of *Small Is Beautiful* is devoted, has caused some confusion. Schumacher was not a Buddhist, and his economics are firmly Christian. Queried about Buddhist economics during his last American lecture tour he conceded, "I might have called it Christian economics, but then no one would have read the book" (*Doing It!* No. 7, July 1977).

[17] Schumacher died of a heart attack on September 5, 1977, while traveling on a train in Switzerland.

with these characteristics, Schumacher pleaded, we could have a revolution in technology which would reverse the destructive trends that threatened us all.

It is instructive to compare this claim with one made by an American industrialist in 1923: "It will be large enough for the family but small enough for the individual to run and care for. It will be constructed of the best materials, by the best men to be hired, after the simplest designs that engineering can devise. But it will be so low in price that no man will be unable to own one—and enjoy with his family the blessing of hours of pleasure in God's great open spaces." This statement bears a striking resemblance to Schumacher's criteria; in fact, it was written by Henry Ford (*My Life and Work,* Garden City, N.Y., 1923), and describes his "universal car," the Ford Model T. Ford, like Schumacher, also claims that this will be a nonviolent technology! He wrote: "When the automobile becomes as common in Europe and Asia as it is in the United States the nations will understand each other. Rulers won't be able to make war. They won't be able to because the people won't let them . . . This is the biggest thing the automobile will accomplish—the elimination of war. The automobile is the product of peace."[18]

Ford was right in some ways. The automobile, which, till his Model T appeared, had been a luxury product whose ownership was restricted to the upper middle classes, did become widely accessible. But Ford was also wrong; or perhaps understandably, unable to see the whole future. The automobile, when multiplied and con-

[18] Interview with Norman Beasley, "Henry Ford Says," *Motor,* January 1924. Ford was, of course, mistaken about the automobile eliminating war. It was the automobile engine which revolutionized warfare and allowed the successful blitzkrieg tactics of World War II.

centrated, is a graphic example that—as regards atmospheric pollution—small is not always beautiful. And Schumacher's claim in his book that "small-scale operations, no matter how numerous, are always less likely to be harmful to the natural environment than large-scale ones" is likewise often untrue. The flush toilet, the throwaway container, and the aerosol can are all "small" devices whose cumulative effect can be environmentally devastating. In the less developed countries, fecal contamination of the environment and deforestation (not by the paper industry but for ordinary household cooking purposes, as in Haiti) are only two examples of the destructive environmental impact of "small" activities.

Ford was also mistaken in his blithe assertion that the automobile was a nonviolent technology. Obviously, the car has done inestimable damage to the countryside as well as to the city, and, not least, to traffic-accident victims. This is not so much to criticize Ford as to point out that it is the concept of nonviolent technology which defies rational analysis. Is a man with a bulldozer more violent than a man with a shovel? Or a hundred men with a hundred shovels? So many of the ideas in *Small Is Beautiful*—nonviolent technology, technology with a human face, even the title itself—are attractive but unproved aphorisms.[19]

The impact of Schumacher, through his book, on that part of Western society that was discontented with modernization, and, to a lesser degree, on some people in the

[19] Similar Schumacherisms, often obscure, abound. "When we consider abnormal that which we now think is normal, that's when we become realistic." Or, "Extreme poverty isn't caused by lack of money or hardware, it's mental." And the cryptic "All you need is simple materials. The Taj Mahal wasn't built with Portland cement" (New York *Times*, October 26, 1975). No, it was built with marble.

underdeveloped countries who were afraid of what modernization would do to their traditional societies, was immediate.

One Schumacher statement formed the *a priori* assumptions of the nascent movement: "Each particular type of technology is itself a political and social shaping agent, with far-reaching sociological consequences."[20] This was a restatement of the media-guru Marshall McLuhan's claim that "any technology gradually creates a totally new human environment," the basic message of his *Understanding Media: The Extensions of Man* (Toronto, 1964). Both McLuhan and Schumacher offered little documentary proof for their assertions. An examination of history, however brief, casts doubt on their theory of technological determinism.

No one would question that modernization, chiefly characterized by industrialization, has shaped the contemporary environment. This was pointed out as early as 1948 by Siegfried Giedion in *Mechanization Takes Command* (New York, 1948). Giedion, a Swiss architectural historian, in a more thorough way than anyone since, analyzed the process of mechanization as it affected a whole range of activities: industry, agriculture, the home. Although critical, Giedion was essentially optimistic, for since "mechanization is the outcome of a mechanistic conception of the world," it was a result, not a cause, and could be controlled. This is in contrast to McLuhan and Schumacher who maintain that technology has predictable and universal effects *in spite of* the way it is applied.

One area that Giedion did not examine was the military. This was curious (especially in 1948) since the lead-

[20] "Using Intermediate Technologies," in Gwen Bell, ed., *Strategies for Human Settlements: Habitat and Environment* (Honolulu, 1976).

ing edge of technological invention is often related to machines used for war. One would have expected, according to technological determinism, that modern war would differ radically from preindustrial warfare. Though this was superficially true, it was also true that Napoleon's dictum "The morale of an army counts at least two thirds; its organization and equipment are the rest" was still applicable as regards the French, for instance, in 1940, or the Americans in Vietnam in the 1960s and 1970s. During World War II it was not shown that leadership, bravery, or tactics for that matter were any less important than they had been in the preindustrial wars. World War II was certainly infinitely more destructive to the civilian population than had been wars in the eighteenth and nineteenth centuries, but this was the result of political decisions and not caused by technology. The barbarism of World War II was not something new; rather, it was something very old. Paradoxically, the mechanization of warfare reduced military casualties considerably when compared to the slaughter of the 1914–18 war.

One event of World War II that seems to be a direct result of technology, and which is often used as a shameful symbol of the Modern Age, is the Nazi death camp. Without mechanization, and all the chilling rationalization of the whole process, such mass slaughter would be impossible to even contemplate. Surely this would seem to be the ultimate effect of the bureaucratization of the individual and the philosophy of efficiency. But is it? The recent exposure by Alexander Solzhenitsyn of the Soviet *gulag* system, a network of prison camps extending over the entire U.S.S.R., which was initiated before Hitler came to power in Germany in 1934 and continues to this day, was accomplished, if that is the word, with the bare minimum of technology.[21] Where there is a will, there is a way; the evil is in the man, not in the machine.

Television is another technology that is sometimes claimed as the archmanipulator of modern man. It is rapidly becoming a worldwide phenomenon—723 million people watched the Apollo II moon landing in July 1969; probably more watch any year's World Cup soccer final. But is this the homogenization of culture that McLuhan foretold? Not quite.

The Japanese are apparently the world's most compulsive TV viewers, spending as much as half of their leisure time watching the box, yet 85 per cent of the programs are produced locally, and foreign programs (including America's) are not particularly popular. Kenya produces programs for much of black Africa; since violence is not permitted on the screen at all, virtually all American programs are banned. The Soviet Union, which initiated its television in 1939 and has one of the largest systems in the world, shows only local programs, with the exception of sports events. The way that television is used as a direct function of the cultural, economic, and political environment (not vice versa). Cuban television has no foreign programs at all, is state operated, and its featured star is Fidel Castro; Indian and Chinese television is almost completely educational, and sets are public rather than private installations; French television is state controlled and, in effect, an extension of the party in power; Canadian television has two separate government-regulated

---

[21] The *gulag* system, according to Solzhenitsyn description, is very much a low-tech effort: labor intensive and with no capital investment or machinery. For example, during 1931–33 the prisoners built the 140-mile White Sea–Baltic Canal virtually by hand; see A. I. Solzhenitsyn, *The Gulag Archipelago, 1918–1956 (III–IV)*, trans. Thomas P. Whitney, New York, 1975.

networks, one English language and one French language; television in Hong Kong goes to great lengths to avoid offending the neighboring People's Republic of China; whereas between Israel and its Arab neighbors, "peace is war continued on television." There is little evidence that "television is itself a political and social shaping tool."[22]

Just as Marshall McLuhan tended to ignore the specific way in which television was used around the world, and the different impact that technology could have in different cultures, so, too, did Schumacher, writing about technology and development, tend to oversimplify and homogenize the relationship between men and machines.

There was a curious unworldliness to *Small Is Beautiful*. It isolated man, his natural environment, and his machines and ignored history, culture, and politics. When the latter were mentioned, they were as minor factors— "technology must be adapted to local traditions" or "political factors should also be considered." There seemed to be little appreciation that technology is the creation of man, and the latter, with his dual propensity for good and evil, is always facing a choice. His decisions are conditioned, more than is generally admitted, by his history. If there was something dislocated in this book, it may have been due in part to Schumacher himself, the perennial outsider, or to the rootlessness and lack of historical sense of the Californized youth who made up an important part of the AT movement.

## THE AT MOVEMENT

The publication of *Small Is Beautiful* not only made

22 This information is contained in an article by Mary Jean Haley, "World Television," *CoEvolution Quarterly*, Winter, 1977–78.

E. F. Schumacher a public figure, it also brought appropriate technology out of the netherworld of academic conferences and government studies and into view. AT, as it was referred to, became a *cause célèbre* and, rapidly, a protest movement.

In a protest movement it is the protest which predominates, and the root of AT's protest was a discontent with the Modern Age, the remedy for which was seen to be demodernization. In *The Homeless Mind* (New York, 1973), the sociologist Peter Berger and his coauthors have described the way that the Modern Age, whose central feature is technological production, has affected not only man's physical environment but also his consciousness. They isolate two effects in particular. First, there is the "rationalization," via science and technology, of almost every aspect of human experience. This brings about an anonymity in social relations which is often experienced as alienation. The German architect Ludwig Mies van der Rohe recognized this in 1924 when he told his students at the Bauhaus, "We are concerned today with questions of a general nature. The individual is losing significance; his destiny is no longer what interests us. The decisive achievements in all fields are impersonal and their authors are for the most part unknown. They are part of the trend of our time towards anonymity."[23] It is only in the twenties that one finds people who are willing, at least publicly, to accept the impersonality that is implicit in the rationalizations of modern industrialization. This was soon to be replaced by a mawkish sentimentalism and later, by fear.

The second effect that Berger describes derives from the "bureaucratization" of society: procedure, orderli-

[23] "Baukunst und Zeitwille," *Der Querschnitt*, No. 4, 1924.

ness, and the compartmentalization, not only of the means of production but also, as a consequence, of political and social institutions. The most visible aspect of this bureaucratization is sheer size. Though this is hardly new (it was pointed out in *The Castle* by Franz Kafka in 1926) and is often exaggerated, bureaucratization affects the individual's life even more than rationalization, and hence the strong *popular* response to "small is beautiful."[24]

An integral aspect of modernization has been society's "solution" to the inevitable, and increasing, anonymity and depersonalization of public life: the *privatization* of many activities which were previously considered public. This privatization serves as a kind of balance, providing, among other things, the personal identity that is lacking in the public realm. The breadth and depth of the privatization of modern society has been documented by Martin Pawley, an English social critic, in *The Private Future* (London, 1973). This book was attacked by liberals, for not only did it point out that privatization was a *fait accompli* and probably irreversible, but it dared to imply that this was what the majority of people actually wanted.

On the one hand, AT was searching for community lost. Much AT writing concerned itself with cooperatives, communes, neighborhood groups, and general "community." The socialist wing promised redemption from privatization by, paradoxically, even more institutionalization. The traditionalist wing sought salvation in a return to the tyranny of village life.

[24] If anything, life today is less bureaucratic than twenty-five years ago. Antimodernists should also explain why most less developed countries are considerably more bureaucratic than most developed ones.

On the other hand, AT was also an (unwitting) victim
of privatization. Self-sufficiency, do-it-yourself, and decentralization are all tactics for increasing privatization.
Consider this advertisement for one AT handbook: "You
can heat or cool your home, run a shop or factory, operate your car, cook your food—in short, supply *all* your energy requirements—with largely non-polluting, completely
renewable sources of power *that you control* [emphasis in
original].[25] "The purpose of AT," the National Center
for Appropriate Technology in Butte, Montana, stated,
"is to make people more self-reliant."

The ability of AT to maintain a two-faced posture
could be attributed to the virtual absence of self-criticism
within this movement. At a certain time, around 1974,
though this is hard to pinpoint, AT achieved the status of
a True Belief (a phrase originally coined by Eric Hoffer).
This was evidenced by the fact that it became usual to
refer to it as *A*ppropriate *T*echnology, and also to make
explicit references to the AT *movement*. AT publications
became little more than tracts; AT conferences eschewed
any semblance of scientific objectivity. The important
thing was to convince people, to advance the cause. Information which was incomplete in the first instance was
printed and reprinted without ever being verified or
substantiated.[26] The AT response to criticism was similar to that of the earlier antiwar movement: "If you are
not part of the solution, you are part of the problem."

This was not a healthy state of affairs. The French

[25] Mother Earth News, eds., *Handbook of Homemade Power*
(New York, 1974).
[26] Particularly flagrant were the continued references to Communist China's espousal of AT. "Walking on two legs" implies
small industry *and* large, with considerable emphasis on the latter;
see Chapter 3, below.

journalist and writer (*Without Marx or Jesus*) Jean-François Revel has described the situation, though in a
slightly different context: "To hold a debate means ex-
posing arguments, therefore spreading them, and this increase of awareness may in the end prove overpowering.
Better by far to suppress it all, either by direct censorship
for the power is there for that, or by indirect censorship.
The technique of indirect censorship, which every *true
believer* [my emphasis] begins by applying to himself for
himself, means above all the deliberate suppression of arguments and facts put forward by enemies of the faith,
then suspicion of their motives, and, as a last resort, accu-
sations that they are sterile, destructive and having nothing new to propose."[27] AT passed the second stage that
Revel describes quite early, and there are indications that
it had even reached the third.[28]


## THE BANDWAGON

AT was a protest movement and for many it also became a True Belief. But it was more than that; AT developed into a bandwagon of Pullman-car proportions. And
what a strange set of traveling companions one found:
well-dressed World Bank economists rubbing shoulders
with Gandhians in metaphorical, if not actual, dhotis; en-

[27] *La Nouvelle Censure;* translated extracts appeared in the New
York *Times Magazine,* December 11, 1977. Revel is referring here
to the reaction of the so-called Eurocommunists to his book *The
Totalitarian Temptation* (trans. David Hapgood, Garden City,
N.Y., 1977); however, the same observations seem pertinent to
other movements of the left.
[28] Steve Baer (see Chapter 4) recounts that "one guy told me,
after my speech at Amherst last year, that he didn't know if he
and his friends would ever be able to speak to me again, because I
had said bad things about 'Appropriate Technology.'" Interview in
*Solar Age,* January 1978.

vironmentalists, Utopians, and *bricoleurs;* conventional
politicians like President Jimmy Carter and less conventional politicians like Governor Jerry Brown of California, who had both met E. F. Schumacher (himself recently
made a Companion of the British Empire by Queen Elizabeth II). The American National Academy of Sciences
had recognized AT and, more importantly, so had the
United States Congress, with $20 million to finance an
organization called "Appropriate Technology International." It was possible that AT had opened the eyes of
all these passengers; it was much more likely that it was
to be their influence on AT which would be more
significant and lasting.

The largest contingent was from the international economic development community which included the national aid agencies in the industrialized countries, the international agencies of the UN, and the various smaller
volunteer, religious, and international charity groups. The
idea of intermediate technology did not initially receive
overwhelming support from these groups; perhaps they
were put off by the name. The world of the United Nations resembles nothing as much as an English club, and
any implication of second best would be immediately rejected. By the time that AT had become a popular issue,
they were prepared to accept it, but with certain important provisos.

Development in the poor countries means *economic* development. Whether this development follows the Chinese
model, the South Korean model, or the Brazilian model, it
implies, in fact assumes, economic growth. Hence, when
AT was accepted by the international economic development groups it was as a strategy for growth, very much the
sort of thing that Schumacher first suggested. The limits-
to-growth ethic that formed a large part of *Small Is*

*Beautiful* and was an unassailable truth for the AT movement in the United States was firmly rejected, if it was
mentioned at all. The demodernizing and countermodernizing trends that have been previously described as part
of the original AT argument were assiduously ignored.

If the international development community was riding
in the first-class section, there was a large group that had
been on the train first, but was now sitting at the back of
the car—the youth culture.

So much has been written about the youth culture, particularly in the United States, that few would contemplate
the possibility that much of it was fiction. But slightly
more than ten years after *Time* magazine announced "the
year of the hippie" in 1968, it is apparent that much of
what was written about and during that period was cant.
The American youth movement is likely to be shorter-
lived than the German *Jugendkultur,* which in its various
forms lasted about forty years. The changes which
seemed to many to have revolutionary implications have
turned out to be manifestations of fashion.[29] John Lukacs,
in *1945: Year Zero* (Garden City, N.Y., 1978) puts this
period into historical perspective: "These new generations
were playing at revolution, not making it. Eventually they
grew tired of this kind of game. By the 1970's the revolutionary temper subsided, because it was not genuine. All
superficial manifestations to the contrary notwithstanding,
the so-called revolutionary ideas and the radical practices

[29] There were few who sang the praises of the counterculture in
the late sixties who were willing to recognize that it was first and
foremost a consumer phenomenon. In *The Making of a Counter
Culture* (New York, 1969), Theodore Roszak did wonder if the
counterculture could "survive these twin perils: on the one hand,
the weakness of its cultural rapport with the disadvantaged; on the
other its vulnerability to exploitation as an amusing side show of
the swinging society." It couldn't and didn't.

of the new generations were not that different from those
that had been current in Year Zero (1945). In the United
States, for example, the notion that the children of the
1960's or 1970's—many of them the grandchildren of a
generation of flappers—were revolting against a parental
generation of strict, narrow, and Victorian or authori-
tarian manners and standards is not even worthy of the
name of legend: it is a tale profitably told by public idiots,
signifying nothing."

It is precisely because of these "profitable tales" that it
is difficult to disentangle the actual contributions of individuals who are often associated with AT from the mere-
tricious claims of certain self-serving publications. Nevertheless, it is a fact that some of the earliest supporters of
AT were found among the youth culture. The later acceptance of these ideas by the establishment was, to a certain extent, a belated attempt to appear to be "with it."

If the interest in AT of the development economist was
as a low-cost growth strategy, the attraction to the youth
culture was often the hardware. What has been referred to
as a "social" and "sexual revolution" should more properly be called one of "manners," and the attraction of AT
for many youths was as a technological equivalent to
"our" clothes and "our" music.

However, the involvement with gadgets was not only at
the symbolic level. The counterculture, despite its name,
was American culture, and Americans are handymen *par
excellence;* unlike AT in Europe, which was essentially
literary, AT in America belonged above all to the inven-
tor-tinkerer. His frail (unsubsidized) accomplishments,
because unique, became the mainstay of many a bureaucrat, for, like other protest movements, AT could be described as an inverted pyramid—a great deal of verbiage
and speculation resting on few accomplishments.

There was another group that attached itself to this movement in a more opportunistic way: the writers of the neo-Utopian left. AT was seen by them, variously, as a means of achieving agrarian socialism, communalistic socialism (almost indistinguishable in some cases from communistic socialism), and other, vaguer, populist Arcadias.

The neo-Utopians gave rise to a number of hybrids such as radical technology, alternative technology, liberatory technology, as well as Utopian technology. Most of these should not be taken too seriously as terms since they were primarily used to make alluring (to the young) book blurbs: "This is a book about technologies that could help create a less oppressive and more fulfilling society."[30] The term *"alternative* technology" was quite popular for a time (particularly in Britain), as it seemed to embrace alternative life-styles, alternative sources of energy, and various other unstated but implicit options.

Some of the hybrids had little or nothing to do with AT. "Liberatory technology" was coined by Murray Bookchin in *Post-Scarcity Anarchism* (Berkeley, 1971): "To an ever-growing extent, technology is viewed as a demon, imbued with a sinister life of its own, that is likely to mechanize man if it fails to exterminate him. *The deep pessimism this view produces is often as simplistic as the optimism that prevailed in earlier decades* [my emphasis]. There is a very real danger that we will lose our perspective toward technology, that we will neglect its liberatory tendencies, and, worse, submit fatalistically to its use for destructive ends." This was, if anything, a critique of the standard leftist position. Bookchin went on to propose that society could be based on decentralized production, using

[30] Peter Harper et al., eds., *Radical Technology* (New York, 1976).

almost no human labor, oriented toward human needs, and freed from all considerations of profit and loss. Bookchin's Utopia (though more idealistic) was actually closer in many ways to the vision of R. Buckminster Fuller than to that of E. F. Schumacher.

The name of Ivan Illich, a mercurial critic of modernization, was also linked to AT, yet his idea of a "convivial society," though vaguer, was similar to Bookchin's. It is true that Illich, like Schumacher, pointed a finger at technology as the root cause of most social evils. In *Tools for Conviviality* (New York, 1977) he categorized all machines as either "convivial" or "manipulatory"—a distinction which was not clear, but not obviously antimodern: the car was manipulatory, but the telephone accepted as convivial. He did make the absolutely crucial distinction between work and labor: in a convivial society, he claimed, machines (of the convivial variety) would minimize labor while maximizing work. Convoluted thinking like this made Illich frustrating to read, but much more interesting than most of the Utopian catechists.

A thinly veiled Luddism permeated the writing of the neo-Utopians. The Luddites were a kind of antitechnological Ku Klux Klan that operated in the British Midlands during 1811–16. They represent one of the last *popular* (as opposed to elitist) outbreaks of resistance to modernization in the Western world. They were protesting the introduction of mechanized textile equipment which had resulted in unemployment (for the artisan-weavers), lower wages, and, to add insult to injury, lower quality cloth. Their leader was a (mythical?) "General Ludd," and masked groups of Luddites, operating at night, destroyed large quantities of weaving machines. These forays grew into full-scale riots which were aggravated by a post-Napoleonic War depression and a poor harvest. In the event, Luddism came to an end as a result of a combination of harsh military repression and reviving prosperity. The verbal assaults of the modern Luddites such as Illich or Ellul are unlikely to bring about the former; they will probably be done in by the latter.

Critics of AT often raised the issue of appropriateness —"appropriate to what?" In different contexts, AT could mean appropriate to *economic* objectives (labor-intensive), appropriate to *social* objectives (decentralizing), or even appropriate to *political* objectives (demodernizing), and occasionally all three. It could also mean appropriate to the *environment*.

The notion of AT as environmentally benign was present in Schumacher's proposal for a nonviolent technology, and in most of the later hybrids. The most cogent argument for an AT-like approach to energy has been put forward by Amory Lovins, a British scientist and spokesman for Friends of the Earth, Inc., an environmental lobby in the United States, in *Soft Energy Paths* (Cambridge, Mass., 1977). Lovins' goal was to explore the implication of an energy program based on the use of only renewable resources and on the resolute abjuration of nuclear power. He claimed not to be a True Believer, however: "I do not pretend here to neutrality: but not for the reasons some might suppose. If I seem to be presenting advocacy as well as analysis, it is not because I began with a preconceived attachment to a particular ideology about energy or technology, such as the 'small is beautiful' philosophy that some have tried to read into my results. It is instead because the results of the analysis so impressed me." A misunderstanding arose out of Lovins' decision to call his option "soft technology," a term that had been coined by an English writer, Robin Clarke, to describe a Utopian technology that was supposed to respond to a whole range of the by-now-familiar ills of modern society. Lovins' use of "soft technology" referred specifically to *energy* technologies.

The main characteristics of the soft energy approach were a reliance on renewable resources (mainly solar), a decentralization of energy sources (primarily on economic grounds) which allows a matching of scale and energy quality to end-use requirements. The reader should note that this implied small energy plants for small users (domestic) and large plants for large users (industrial). Lovins made a convincing case for decentralization, particularly for domestic uses, but he could not resist embroidering his argument with the assertion that soft technologies will be "easy to understand and use without esoteric skills." This undoubtedly endeared him to the mainstream AT movement, though he wisely skirted the issue of possible conflicts between ease of use and ease of understanding, preferring instead to make oblique references to gardening and do-it-yourself carpentry.[31]

The result of the environmental concerns of AT was a decided bias toward energy-producing technologies, as opposed to manufacturing technologies, particularly after the "energy crisis" of 1973–74. At this point one could observe a difference in national motivations in using soft energy technologies. Countries such as Japan and India, which faced a virtually fossil-fuelless environment, were forced to seek other energy sources. The industrialized countries, which face a polluted environment with the likelihood of further degradation by nuclear wastes, were looking at soft technologies as a nonpolluting source of

[31] In spite of such lapses, *Soft Energy Paths* is by far the most intellectually responsible book dealing with AT. This may be because the author is a physicist or because he is dealing specifically with a defined field, energy.

energy. Finally, there were countries such as the United States, whose stated geopolitical aim was energy self-reliance and whose interest in renewable sources stemmed from this concern.

The three latter groups of passengers—the youth culture, the neo-Utopians, and the environmentalists—are all situated in the industrialized countries, and all share a common belief: in one way or another, there has to be a limitation, a slowdown, or a halt to economic growth—what I have previously referred to as "demodernization." A full discussion of the limits-to-growth position is beyond the scope of this book, but the relationship between this position and AT is worth noting, since it was neither self-evident nor always consistent.[32] This inconsistency was most obvious to the less developed countries, whose rejection of no-growth was usually unambiguous.

There was a final ideology on the bandwagon that was neither particularly fashionable nor recent: the Gandhian philosophy. It was Mohandas Gandhi who coined the first AT epigram, "Production by the masses, not mass production." It was he who popularized the first AT device: the *charkha*, or spinning wheel. It was he also who voiced the traditionalist critique of modernization: "The traditional old implements, the plough and the spinning wheel, have made our wisdom and welfare. We must gradually

[32] The cult of demodernization has no lack of advocates; indeed it is so fashionable that it hardly needs any substantiation at all. Critics of antigrowth are few and far between. The adventurous reader is directed to *In Defence of Economic Growth* (London, 1974), by Wilfred Beckerman of University College London; *The Doomsday Syndrome* (New York, 1972) by John Maddox; an impassioned attack on "the professors of apocalyptic holocaustology" by Petr Beckmann, *Eco-Hysterics and the Technophobes* (Boulder, Col., 1973); and Herman Kahn's *World Economic Development* (New York, 1979).

return to the old simplicity . . . I do not believe that multiplication of wants and machinery contrived to supply them is taking the world a step nearer the goal . . . India's salvation consists in unlearning what she has learnt in the last fifty years. The railways, telegraphs, hospitals, lawyers, doctors, and such like all have to go; and the so-called upper classes have to learn consciously, religiously and deliberately the simple peasant life, knowing it to be a life giving true happiness . . . You cannot build non-violence on factory civilization; but you can build it on self-contained villages."[33] It was Gandhi who, before China's Mao Tse-tung, recognized that the peasants should be the basis for economic development in Asia.

The debt to Gandhi is sometimes acknowledged (Schumacher referred to him as "the greatest economist of the 20th century"[34]), and is sometimes implicit, but there is no doubt that Gandhism has been a powerful ideological influence on the AT movement.

The difficulty with discussing Gandhism is that it is an economic theory founded on the teachings of a man who was a saint. References to his pronouncements are like references to the Christian Bible; the reference itself is the proof. But Gandhi was also a political leader and a politician, and his ideas can be assessed in a particular historical context. His campaign of *satyagraha,* passive resistance and noncooperation, was so successful against the British precisely because it was in fact coercive, and it was seen by most of his followers as a political tactic rather than as a moral principle. His disciple Jawaharlal Nehru wrote in 1929: "The great majority of us, I take it,

[33] Quoted by Sir Penderel Moon in *Gandhi and the Modern India* (New York, 1969).
[34] Interview in the New York *Times,* October 26, 1975.

judge the issue not on moral but on practical grounds, and if we reject the way of violence it is because it promises no substantial results." Likewise the Gandhian philosophy of the self-contained village and the simple life was not taken seriously by the majority of Indian politicians even during Gandhi's lifetime, with the result that immediately after independence, India embarked on a modernization program. The *khadi* (homespun cotton) and village industries movement "still exacts from an unbelieving Indian government tribute in the form of large grants" but has little actual influence on Indian policy makers.[35]

Of course, none of this necessarily disproves the validity of Gandhi's approach, but it does indicate the problems of the traditionalist when faced by modernization. The moment that the traditionalist incorporates modernizing tendencies (in Gandhi's case, nationalism, land reform, democratization of the caste system), the process of "contamination" begins, and the further modernization of the modified traditional society, if not inevitable, is at least likely.[36] Thus the pressures on India the nation-state eventually resulted in the fabrication of atomic devices, something the government of India had resisted for a long time, partially, at least, in deference to nonviolence. The organization of Indian industry along decidedly non-Gandhian lines was a similar side effect which is unlikely to be redirected, in spite of the pronouncements of the Janata party, elected to power in 1977.

In India, village technology and Gandhism were synonymous. Gandhi himself established the All-India Village Industries Association, whose center was in Wardha

[35] Sir Penderel Moon, op. cit.
[36] Nehru has been quoted as remarking, "We [India] have atomic energy, and we also use cow dung." It is not recorded if he was being poignant or merely sardonic.

where Gandhi lived and worked in the late 1930s. This movement continues in present-day India, though for a long period it was vastly overshadowed by Nehru's series of five-year plans. The rise of AT in the West gave new encouragement to the previously ignored advocates of village and rural technology. An Appropriate Technology Development Association was formed, situated, aptly enough, in the Mahatma Gandhi Building in Lucknow.

But the Gandhians were no recent converts, and so it is not surprising to find a degree of skepticism in their views on "imported" AT. The head of the Indian AT association, M. M. Hoda, wrote in alarm in his first newsletter of July 1977: "Some people now claim that the sophisticated western technology is the most appropriate technology. On the other hand, one school of thought suggests that appropriate technology should be used only as a stop-gap arrangement in those fields where it is not possible to import most sophisticated western technology. Similarly a representative from a major international organization wanted that highly trained and sophisticated engineers should visit the rural areas of the developing countries and design technology to fit into their circumstances. Disaster may follow if multi-national organizations take up to produce small machines in stainless steel packages for the rural areas of the developing countries. This would be the end of appropriate technology."

Is that what AT holds in store for the Third World?

*Chapter 2*

# MILLSTONE . . .

"There is only one way by which poverty in the developing countries can be attacked successfully, and that is by producing more in those nations. In no one of these countries can human needs be satisfied by the simple redistribution of existing income and wealth. In these countries small is not beautiful."

—ROBERT S. MCNAMARA,
*President of the World Bank*

It has been suggested that the less developed countries need an alternative technology to that of the developed countries. This proposal is based on two assumptions: that the world is split into two camps, the rich and the poor countries; and that the technology from one of these camps, the rich, is unsuitable to the needs of the other, the poor.

## DEVELOPED, MORE OR LESS

The key assumption is the first one cited above. If, in fact, the world is radically split in two, then it may be likely that technology and development will occur in two radically different ways. The terminology that has come into use—rich/poor, developed/underdeveloped, East/

West, or North/South—reinforces this assumption. Close observation of the situation does not.

The World Bank published, in 1976, a world atlas that listed population, per capita production, and growth rates of 187 countries and territories. These figures, general as they are, are nevertheless instructive. They indicate, for instance, that the world is not divided into two as regards gross national product (GNP) per person. That of Bangladesh is lowest, that of Kenya is twice as large. El Salvador's is twice as large as Kenya's and the People's Republic of China's is twice that of El Salvador. That of Argentina is twice as large as that of the People's Republic of China. Czechoslovakia's is twice Argentina's, and, finally, that of the United States is double that of Czechoslovakia. I include this rather tedious litany to indicate that, to the extent that GNP per capita represents development (which it does only very roughly), then development can be said to affect *all* countries. Rather than development and underdevelopment, it is more accurate to speak of a *gradient* of development—less developed and more developed. The measure of GNP per capita is, rather inaccurately, a measure of the wealth of *people* (though not of income distribution). The people of one of the superpowers, the Soviet Union, are poorer than those of Spain. The people of European countries such as Yugoslavia, Romania, Bulgaria, Greece, and Hungary are poorer than the citizens of Puerto Rico or Singapore and, more predictably, than those of Libya or Saudi Arabia. But the citizen of Panama is twice as "rich" as his counterpart in Albania. On the other hand, the wealth of *states* (or gross GNP) reflects the over-all power of a country. Not surprisingly the two wealthiest states are the United States and the Soviet Union. The People's Republic of China ranks sixth. Another "less developed" state, Brazil, is tenth. The twenty richest states in the world include Mexico and India, as well as countries with relatively low GNPs per capita such as Poland and Spain. On that list, not one of the supposedly rich OPEC countries appears in the top twenty.

There are other anomalies that do not support the thesis of a polarized world. India, which has one of the lowest per capita incomes in the world has the seventh largest steel industry in the world. Though there are many less developed countries that depend on exporting raw materials, the largest exporters are the more developed countries such as Canada, the United States, and the Soviet Union. Is the primacy of agriculture a measure of underdevelopment? Hardly, if one considers Canada or Holland. Is exploitation by the multinationals the cause of poverty? What, then, about China and India, where there are virtually no foreign-owned industries?

My point is not that there is no inequality, or even injustice, in the world; there is. But the model of a world split into two, the rich and the poor, is simplistic and does not reflect the realities of countries in various stages of development.[1] Neither does it reflect the political realities of the power of states, and it certainly does not reflect the internal realities of the division of wealth within countries.

[1] Professor P. T. Bauer, of the London School of Economics, has written critically of current attitudes toward development and the less developed countries: "It is a travesty, and not a useful simplification, to lump together Chinese merchants of Southeast Asia, Indonesian peasants, Indian villagers, tribal societies of Africa, oil-rich Arabs of the Middle East, aborigines and desert peoples, inhabitants of huge cities in India, Africa and Latin America—to envisage them all as a low-level uniform mass, a collectivity which moreover is regarded as no more than a copy of Western man, only poorer, and with even this difference the result only of Western responsibility" ("Foreign Aid Forever?" *Encounter,* March, 1974).

If the world cannot be neatly parceled in two, what about technology?

Though the term "technology" is used in the broadest possible way, especially by its critics, it actually has little meaning unless it is used in a specific context. Even the term *Modern* technology" is virtually meaningless; for instance, in a "modern" Western city one finds, contemporaneously, public transport (the first omnibus service in London was initiated in 1829); water and sewer systems, another nineteenth-century idea and often, in fact, nineteenth-century installations; the automobile, a technology that has been dominant since the 1930s; various networks based on cable, and wireless, transmission; a proliferation of electronic devices that have appeared since the 1960s; and an increasing use of very advanced technologies such as computers, lasers, and semiconductors. Though all of these are identified as "modern," the fact is that technologies in various states of evolution exist side by side.

Technological development advances by fits and starts; sometimes one field is affected, sometimes another. Although science can discard outmoded theories when new ones are improved, technology cannot. Technologically consistent worlds exist only in science fiction novels; the real world is always technologically inconsistent. For this reason it is pointless, and misleading, to describe technology as if it were a national attribute; if it appears to be so, that is only a circumstantial fact. The microscope was invented by a Dutchman and an Italian in the sixteenth century; the steam engine was invented by an Englishman; the cotton gin by an American; the electric battery by an Italian; motion pictures were developed in France, England, and the United States, though today India produces the greatest number of films; the X-ray was invented by a German. Technologies tend to be developed in response to needs and, more importantly, means. No one would refer to the radiotelegraph as a Latin technology, nor to dirigibles as Teutonic technology; it is equally misleading to refer to a "Western" technology. Technologies are "Western" only to the extent that circumstances in the countries of the so-called Western world, which industrialized earliest, prompted and permitted the invention and production of many industrialized technologies. As these circumstances are found in more and more countries, one would expect to find the development of modern technology taking place around the world. This is already happening in various fields and will increase in the future. To speak of "Western" technology being appropriate or inappropriate to the less developed countries is to exhibit a chauvinism that has little foundation in reality. In spite of their sometime rhetoric, the less developed countries understand this better than their well-meaning advocates in the developed world.

The application of technology in the industrialized countries has never been as homogeneous as critics would have us believe. Public transport technology varies significantly between various countries, the result of urban patterns and climate. There has always been a significant difference between the automobile technologies of Europe and North America, based largely on distances traveled and function. Building technology is quite different in Canada from building technology in England, partly because of climate and the wide availability of wood. In most of these cases, the differences in geography, resources, and living patterns have affected the specific technological solutions. It is very likely that the less developed countries will likewise have to modify various technologies to reflect local differences, particularly since, in certain cases, these differences are striking.

## THE LATECOMERS

It is sometimes stated that the less developed countries can take advantage of being "latecomers" in industrialization. According to this argument, originally advanced by Leon Trotsky in *The History of the Russian Revolution* (1932), they can leapfrog those countries that industrialized in the nineteenth century by adopting the most modern machinery. There is a certain amount of truth in this, as the development of manufacturing industry in Brazil, Taiwan, and South Korea shows. However, the latecomer theory ignores one extremely important set of factors. As the Swedish sociologist Gunnar Myrdal has pointed out in *The Challenge of World Poverty* (New York, 1970), most of the less developed countries face a different situation from that which existed in the advanced countries when they began their development, most of them in the 1850s. The main differences are climatic (tropical versus temperate), population (very large versus fairly small), and in availability of resources (water, fossil fuels, arable land). There are also significant differences in the evolution of cultural and political institutions. Religious reformations, the development of scientific thought, and the consolidation of the nation-state were in a more advanced stage when the European countries began their greatest industrial development in the 1850s than they are today in many of the less developed countries. Because of these differences, for many (not all) of the less developed countries, being latecomers is no advantage at all.

One example of how the different conditions in less developed countries require a different technological response than in the advanced countries is the problem of urban sanitation. Urbanization has almost always been

proportional to the wealth of a country; the richer the
country, the smaller the rural fraction of the population.
This is still the case; a poor country such as India has less
than a tenth of its population living in cities, while almost
three quarters of relatively richer Chile live in cities and
towns. What has changed in the last twenty years is the
actual size of the cities themselves: although India is not
an urbanized country, it does have two cities of over 5
million people. Mexico City, which is now, according to
the United Nations, the second largest city in the world, is
growing at the rate of 1,000 persons per day. The provision of basic urban services such as water and sanitation
in these cities has not been able to keep up with this
growth, and, as a result, half of the urban population in
the less developed countries do not have domestic piped
water, one quarter do not have sanitary facilities of any
kind, and only one quarter have access to facilities that
are connected to urban sewers. Why have underground
sewers, a "Western" technology, not been able to solve
this important problem?

The first answer is an economic one—the less developed
countries (with rare exceptions) are poor and do not have
the necessary capital to invest in underground sewers,
which are not cheap. But the bottleneck is not only monetary. A sewer system requires water; and half of the urban
population does not have indoor plumbing. Many developing countries situated in the tropical and subtropical
regions find themselves with periodic or permanent short-
ages of water, both for agricultural and domestic uses. In
the temperate regions, where sewers originated, water was
plentiful so it was not surprising that it was used as the
transporting medium for human wastes. The rapid urbanization of cities in the less developed countries likewise complicates the picture. When sewer systems were

built in Hamburg (1842), Brooklyn (1857), and Paris (1860), these cities all had populations of less than one million. Today it is necessary to provide sanitation for cities six or seven times that size. The scale of the problem faced by the less developed countries, in this case, is quite staggering. The financial resources are minimal, and the conventional solution (i.e., sewers), by all indications, is inadequate. It is also important to appreciate that most households in European and American cities were already provided with running water before sewers were built, and in fact the latter were a direct result of the former. This is not the case in most of the less developed countries.

I will leave aside, for the moment, the question of what kind of options the developing countries actually have in dealing with urban sanitation. The main point is that the technological solution which had been previously adopted by the advanced countries has not been able to solve the problem in the less developed countries. Thus this seems to be one case where an alternative solution will probably have to be found. To this extent it is correct to say that the poor countries do sometimes need different technological solutions than those which have been previously developed in the industrial countries. But are these different technologies the result of a new and different approach?

It is first necessary to differentiate between growth and development, which are certainly not the same thing—the former does not guarantee the latter. If the goals of growth are primarily economic, those of development are more broadly social; to put it simplistically, growth affects production, development affects distribution. While growth does not guarantee development (the case of oil-rich Nigeria comes to mind), the corollary is also true:

without growth in the less developed countries, development will be difficult and in most cases impossible.

The point has already been made that in the less developed countries one of the appeals of a return to indigenous technology has been as a countermodernizing influence. It must now be said that the extent of this countermodernization is extremely limited. There are indications that in some countries of the Middle East, recently Iran[2] and Turkey, there have been popular movements based, at least in part, on a turning away from Western modernization back toward traditional Islamic life; on the other hand, the otherwise traditional states of the Arabian peninsula have had no difficulty in combining modernization with extreme religious orthodoxy. Another isolated case is India, where the government of Prime Minister Morarji Desai, following its election in 1977, made specific references to a return to Gandhian principles. Pakistan has made references to "Muslim socialism" though it is not clear what this entails. Like the African nationalist movements, it seems to be a posture which does not call into question modernization, only its superficial European appendages.[3]

It is important to note that true antimodernism in the less developed countries emanates from the right—either from simple xenophobia (usually religious) or from a

[2] The forty-year rule of Mohammad Reza Pahlavi, the Shah of Iran, was terminated in January 1979, in part by an uprising of conservative Shia Muslims led by the Ayatollah Ruhollah Khomeini; one of their grievances was the *gharb-zadegi* (Western toxication) that had overtaken Iran as a result of modernization.

[3] The return to traditional dress that was sometimes espoused by Presidents Mobutu Sese Seko of Zaire, Ferdinand Marcos of the Philippines, and former President Luis Echeverria Alvarez, of Mexico was simply a resistance to Europeanization, in this case of clothing, which had nothing to do with antimodernization. It should not be confused with Gandhi's wearing of the dhoti, which symbolized a revolt against modernization *per se*.

more sinister desire to preserve a situation of privilege for a minority group. There is no doubt that the position, and way of life, of the upper middle classes in the less developed countries are considerably more comfortable than those of their counterparts in the advanced countries. The availability of cheap labor is most visible in the existence of servants, a phenomenon which, since 1945, has virtually disappeared from America and Europe. The social position of this group in less developed countries (every middle-level executive has a chauffeur) likewise resembles that of the upper middle class in pre-World War I Europe; this is all now threatened by a general rise in incomes, a growth of a lower middle class, and above all, a general democratization of society.

There is a final nail to be put into the coffin of counter-modernization before it is put to rest. For almost all of the less developed countries, a reversion to a traditional society is an impossibility; *they have passed the point of no return.* There is graphic evidence of this in countries such as Zambia or Ghana, where civil unrest has followed government cutbacks on wages or restrictions on standards of living. Similarly vociferous popular reaction to state-inspired demodernization has occurred in the past in Poland and Peru. As soon as development is set as a national goal, something which has happened some years ago in virtually *all* countries, modernization ideals must be accepted. Their acceptance is reinforced by the fact that pressures of population (present and future) require rapid development in order to avoid economic stagnation or regression. When there is a conflict between modernization ideals and traditional values, it is the latter that will, in many cases, be modified. The postulation of some that each country will develop following its traditions is not necessarily untrue, but it is misleading. Each country

will develop following modernization ideals and will re-
tain many traditions, some of which are neutral and some
of which actually promote development; but those traditions that conflict with the ideals of modernization will
tend to be changed. Once again I must emphasize that this
does not lead to a homogenization of cultures, though it
undoubtedly results in changes, and drastic ones at that.[4]
   The fact is that almost all the less developed countries
have formulated their development objectives *in relation
to the more advanced countries.* This could mean slavish
imitation; in practice, it tends to mean a rejection of certain aspects and acceptance of others. China today is obviously accepting advanced (often Japanese) technology,
but equally obviously is following a socialist road in internal organization. The oil-rich states of the Middle East
are likewise developing selectively. Observers have often
noted that Japan, though it seems to accept many West-
ernisms, does so in a uniquely Japanese way. The important thing to note about this "role-model" type of development is that it enables the less developed country to
learn from the advanced country, to avoid certain pitfalls,
and to take advantage of recent advances in both science
and technology. The ultimate decision remains that of
each particular country.

   [4] There are numerous examples of such changes. The attempt in
Communist Cuba to discourage the attitude commonly called *ma-
chismo,* one of the less attractive characteristics of Latin culture;
the change in the status of women in the conservative Arab states,
such as Kuwait, is only beginning, but promises to alter these cultures beyond recognition; China and Vietnam are both reforming
their writing—Roman letters are used for phonetic transcriptions of
Chinese characters.

## DON'T DO AS I DO

Some theorists propose a new approach to replace the traditional imitative type of development. They suggest that each country should develop according to its own particularities: different cultural backgrounds must be respected; technologies must be adapted to meet specific national differences; people must not have to adapt to technology, but vice versa. According to this view, imported technology represents external cultural domination; this is not only condescending with regard to the less developed countries, it is also erroneous.

Is it possible for a less developed country to turn away from the model of modernization that the more advanced nations offer, particularly if one interprets "more advanced" as signifying not only the nations of western Europe and the United States, but also Japan, the Soviet Union, and, increasingly, Brazil, Venezuela, or Mexico? From a Haitian perspective, the more advanced country might well be Jamaica; from a Chinese, Yugoslavia. The point is, that since the world is not divided into two camps, as many European and American and some Third World critics claim, "more advanced" simply means *any* country which is further along in the development process. The Central American republics import technology from Mexico, Mexico from Spain, Spain from Germany, and Germany from the United States. The exchanged experiences between countries that are close on the development ladder are more likely to be useful than those of countries that are far apart.[5]

[5] "But as countries such as India work out their problems of popular education, birth control, and land consolidation, this experience will be exceedingly valuable to those that follow along the

The view that countries can and should develop according to their own particularities has been characterized by Myrdal as reminiscent of the old, static, anthropological view toward backward cultures, which regarded any change as a "disturbance." It assumed that development could take place without major cultural changes. Although it is generally accepted that popular education, emancipation of women, and general democratization (to name but a few social reforms) are required for development to take place, there are traditional values that oppose these and likely will have to change, if not disappear. Restrictive religious practices, tribal divisions, or traditional elites may contradict modernization ideas (as Myrdal calls them). Whether these obstacles, if indeed they are such, are removed or maintained is of course a national decision, but to start with the assumption that they will not change may preclude development altogether.

The second drawback to the inward-looking mode of development is equally formidable. If a country is to develop according to its particularities, which particularity should technology respond to? Climate, geography, politics, or culture? How does one judge which cultural particularity is fixed and which likely to disappear? Religion? There may be half a dozen religions. How are priorities assigned? Entrepreneurs in many less developed countries belong to racial minorities: their particularities may be different than those of the majority. What about political particularities?[6] How does one respond to the particu-

line. I venture to think that India can often be a better teacher here than the United States. She has been much closer to the practical problem." John Kenneth Galbraith, *Economic Development* (Cambridge, Mass., 1962).

[6] I have been told by the director of one AT group that a project in a Southeast Asian country could only proceed under a par-

larities of a regressive, narrow-minded, kakistocracy?[7]

The examples of countries that have adopted the development-according-to-particularities approach are few and far between—perhaps Japan in the nineteenth century or the People's Republic of China during the so-called Cultural Revolution (1966–69). Significantly, both countries later turned to a more conventional outward-looking modernization. The communist Khmer Rouge regime in Cambodia, in the mid-1970s, embarked on a course of inward-looking development; in the light of the experience of China's Great Leap Forward (under way in 1958), one could have expected the experiment to have ended in agricultural decline and industrial decay. The majority of countries, whether socialist, communist, or capitalist, are pursuing their development according to the role-model approach, though, of course, not all according to the same model.

At this point I should differentiate between two ways in which a less developed country might view technology and development. First, technology may be seen as a stepping stone, very much the way that China has used small industries to bridge the gap to larger and continuing industrialization. It is likely that this approach is a useful one for many countries, particularly as it was also the approach used by many of the advanced countries themselves. The intermediate technology tactic will only be successful if it is clearly understood as such—a tactic in the

---

ticular stricture—it had to take place on land belonging to a member of the President's family.

[7] AT projects have been documented during President François Duvalier's regime in Haiti (1957–71), Shah Mohammad Reza Pahlavi's rule in Iran (1941–79), and General Kjell Laugerud Garcia's dictatorship in Guatemala (1974–78). The lack of real government interest in such initiatives in these nations has made "appropriateness" in technology a dubious goal.

# THE APPROPRIATE TECHNOLOGY MOVEMENT

The first Appropriate Technologist, Mohandas Gandhi, was photographed in 1930 with the precursor of the first AT device,

1

the traditional Hindu spinning wheel (1). Like other examples of AT, the spinning wheel had a symbolic as well as a practical function, being incorporated into the design of the flag of the National Congress Party of India.

E. F. Schumacher, the late economist, was an effective spokesman for AT. He is shown here in the Oval Office of the White House in May 1977 with President Jimmy Carter, who holds a copy of *Small Is Beautiful* (2).

2

The Appropriate Technology bandwagon carries many passengers: "radical" technology is seen as a political tool, as in this 1960s pro-bicycling British poster (3); the environmentalists

THE VC BIKE VERSUS
THE USAF PHANTOM

THE BIKE IS WINNING IN VIETNAM
IT CAN WIN IN LONDON

3

consider "soft" technology as an alternative to centralized industrialization; the neo-Utopian Left calls for autonomous homesteads in a postindustrial Arcadia (4).

4

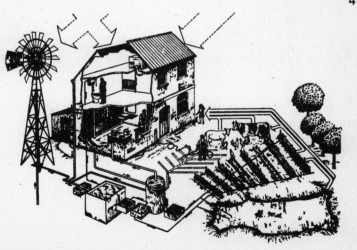

5

Technologically consistent worlds exist only in science fiction
novels; the real world is always technologically inconsistent,
the Third World particularly so. These African Muslims in
Mali have descended from their Japanese bicycles and mopeds

6

for the great Thursday prayer. Electric transmission lines pass
over the mosque, which, for all its elegance, is built out of mud
(5). Technology is inevitably adapted to meet local conditions.
Movable brick kilns are used to fire hand-pressed bricks in

India (6), whereas in Denmark, though bricks are likewise used in construction, entire wall panels are prefabricated in the factory in order to reduce on-site labor (7).

7

## WALKING ON TWO LEGS

The communist Chinese approach to technology is now pragmatic rather than ideological. The forced establishment of small industries based on the "reeducation" of intellectuals and

8

bureaucrats during the Cultural Revolution was an example of an alternative approach to technology, but one which was unsuccessful and has been largely discontinued. It has given way to a blend of technologies along the Western model: precast concrete elements are used for a shipbuilding warehouse in Shanghai (8), whereas in Tachai, terraced housing combines traditional and modern building techniques (9).

9

# CALIFORNIA DREAMING

Buckminster Fuller in the late 1960s (10).

10

"We heard R. Buckminster Fuller lecture in Boulder, Colorado,
and decided to build domes." Many of the Drop City domes
were constructed from used-car tops, and one of them incorporated a solar heater — this, five years before the energy crisis

11

of 1973–74 (11). One of the Drop City builders, Steve Baer
(12), organized the New Mexico desert conference in 1969,
which brought together a number of people from California

12

13

and the Southwest, including Lloyd Kahn (13), future editor of *Domebook,* shown here addressing the meeting in a large geodesic dome.

The entire first page of the fall 1969 *Whole Earth Catalog* was devoted to Fuller (14): "The insights of Buckminster Fuller initiated this Catalog." The influence of publications such as *The Whole Earth Catalog* has not been confined to the United States. In Mexico, Luis Lesur is preparing a series of do-it-yourself books on a wide variety of rural subjects (15).

# Whole Systems

**Buckminster Fuller**

*The insights of Buckminster Fuller initiated this Catalog.*

*Among his books listed here, his most recent is probably the best introduction—it's a succinct summary of what's been on his mind for many a year and what's on his mind now: how mankind may hatch and survive the hatching. An Operating Manual for Spaceship Earth.*

*Of the other, larger, books, Nine Chains to the Moon is his earliest and most openly metaphysical, The Unfinished Epic of Industrialization the most beautiful, Ideas and Integrities his most personal, No More Secondhand God the most generalized, World Design Science Decade (co-authored with John McHale) the most programmatic.*

*People who beef about Fuller mainly complain about his repetition—the same ideas again and again, it's embarrassing. It is embarrassing, also illuminating, because the same notions take on different uses when re-approached from different angles or with different contexts. Fuller's lectures have a raga quality of rich nonlinear endless improvisation full of convergent surprises.*

*Some are put off by his languages, which makes demands on your head like suddenly discovering an extra engine in your car—if you don't let it drive you faster, it'll drag you. Fuller won't wait. He spent two years silent after illusory language got him in trouble, and he returned to human communication with a redesigned instrument.*

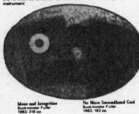

**Ideas and Integrities**
Buckminster Fuller
1963; 318 pp.

**$1.95** postpaid

from:
Collier Books
The Macmillan Company
Order Dept
Front and Brown Streets
Riverside, N. J. 08075

or WHOLE EARTH CATALOG

You belong to the universe. The significance of you will forever remain obscure to you, but you may assume that you are fulfilling your significance if you apply yourself to converting all your experiences to highest advantage of others. You and all men are here for the sake of other men.

I define "synergy" as follows: Synergy is the unique behavior of whole systems, unpredicted by behavior of their respective sub-systems' events.
[Ideas and Integrities]

Thinking is a putting-aside, rather then a putting-in discipline, e.g., putting aside the tall grasses in order to isolate the trail into informative viewability. Thinking is FM—frequency modulation—for it results in tuning-out of irrelevencies as a result of definitive resolution of the exclusively tuned-in or accepted feed-back messages' pattern differentiability.
["Omnidirectional Halo" No More Secondhand God]

*The World Design Science Decade documents contain some that is in the other books and much that isn't. The 6 volume set costs $10.50 postpaid to students (formal and informal); $30.00 postpaid to others. This is a very good deal.*

Order from:
World Resources Inventory Office
P. O. Box 909
Carbondale, Illinois 62901

Man is so deeply conditioned in his reflexes by his millenniums of slave functioning that he has too many inferiority complexes to yield to politikbod reformation. The obsolete genius will be abandoned only when realistic, happier and more interesting games come along to displace the obsolete genes.
[WDSD Document 3]

We find that original question nature is a consciousness of interferences, whether in the computer or the human brain. We find then that original questions are second derivative events in the computer life. [WDSD Document 3]

My Recommendations for a Curriculum of Design Science:
1. Synergetics
2. General Systems Theory
3. Theory of Games (Von Neumann)
4. Chemistry and Physics
5. Topology, Projective Geometry
6. Cybernetics
7. Communications
8. Meteorology
9. Geology
10. Biology
11. Sciences of Energy
12. Political Geography
13. Ergonomics
14. Production Engineering
[WDSD Document 5]

**Operating Manual for Spaceship Earth**
Buckminster Fuller
1969; 143 pp.

**$4.25** postpaid

from:
Southern Illinois University Press
600 West Grand
Carbondale, Illinois 62903

or WHOLE EARTH CATALOG

For $4.95 you can get a paperback called Environment and Change which has an identical "Operating Manual" along with 24 other futuristic articles, including fine pieces by R. G. H. Siu, John R. Platt, Herman Kahn, Robert Theobald, Gunnar Myrdal, David Buzelon, and John Turner.

from
University of Indiana Press
P.O. Box 369
Bloomington, Indiana 47401

or WHOLE EARTH CATALOG

**Nine Chains to the Moon**
Buckminster Fuller
1938, 1963, 375 pp.

**$2.45** postpaid

from:
Southern Illinois University Press
600 West Grand
Carbondale, Illinois 62903

or WHOLE EARTH CATALOG

Common to all such "human" mechanisms—and without which they are imbecile contraptions—is their guidance by a phantom captain.

This phantom captain has neither weight nor sensorial tangibility, as has often been scientifically proven by careful weighing operations at the moment of abandonment of the ship by the phantom captain, i.e., at the instant of "death." He may be likened to the variant of polarity dominance in our bipolar electric world which, when balanced and unit, vanishes as abstract unity 1 or 0. With the phantom captain's departure, the mechanism becomes inoperative and very quickly disintegrates into basic chemical elements.

An illuminating rationalization indicated that captains—being phantom, abstract, infinite, and bound to other captains by a bond of understanding as proven by their recognition of each other's signals and the meaning thereof by reference to a common direction toward "perfect"—are not only all related, but are one and the same captain. Mathematically, since characteristics of unity exist, they cannot be non-identical.

Since Yogi is a personalized art, the art dies with the person. The abstract power involved remains as real and true, always, but it cannot be made utilizable in increasing continuity for the world in general. Christ and his counterparts realized this and were unique in their refusal to apply this power to self ends. It was this personal limitation of the Yogi art which led the gnostic philosophers to search further. They sought a means of limitless articulation.
[Nine Chains to the Moon]

The will of history reads "for everybody or for nobody," and since we balk at "for nobody" it has to be "for every-body." And that's the way it is going, lickety-split and the world around.
[WDSD Document 3]

Personal Self Disciplines, In 1927 I gave up forever the general economic dictum of society, i.e., that every individual who wants to survive must earn a living. I substituted, therefore, the finding made in concept one, i.e., the individual's antrentropic responsibility in universe I sought for the tasks that needed to be done that no one else was doing or attempting to do, which if done would physically and economically advantage society and eliminate pain.
[WDSD Document 5]

Reshape environment, don't try to reshape man.
[WDSD Document 1]

**No More Secondhand God**
Buckminster Fuller
1963; 163 pp.

**$2.25** postpaid

from:
Southern Illinois University Press
600 West Grand
Carbondale, Illinois 62903

or WHOLE EARTH CATALOG

To begin our position-fixing aboard our Spaceship Earth we must first acknowledge that the abundance of immediately consumable, obviously desirable or utterly essential resources have been sufficient until now to allow us to carry on despite our ignorance. Being eventually exhaustible and spoilable, they have been adequate only up to the critical moment. This cushion-for-error of humanity's survival and growth up to now was apparently provided just as a bird meets of the egg is provided with liquid nutriment to develop it to a certain point. But then by design the nutriment is exhausted at just the time when the chick is large enough to be able to locomote on its own legs. And so as the chick packs at the shell seeking more nutriment it inadvertently breaks open the shell. Stepping forth from its initial sanctuary, the young bird must now forage on its own legs and wings to discover the next phase of its regenerative sustenance.

Brain deals exclusively with the physical, and mind exclusively with the metaphysical. Wealth is the progressive mastery of matter by mind.

A new, physically uncompromised, metaphysical initiative of unbiased integrity could unify all humanity. It could and probably will be provided by the utterly impersonal problem solutions of the computers.

Possession is becoming progressively burdensome and wasteful and therefore obsolete.

You and I are inherently different and complementary. Together we average as zero—that is, as eternity.
[Operating Manual for Spaceship Earth]

**The Unfinished Epic of Industrialization**
Buckminster Fuller
1963; 227 pp.

**$4.50** postpaid

from:
Small Publishers Company
276 Park Avenue South
New York, N.Y. 10010

or WHOLE EARTH CATALOG

However,
man unconcernedly sorting mail on an express train
with unuttered faith that
the engineer is competent,
that the switchmen are not asleep,
that the track walkers are doing their job,
that the technologists
who designed the train and the rails
know their stuff,
that the thousands of others
whom he may never know by face or name
are collecting tariffs,
paying for repairs,
and so handling assets
that he will be paid a week from today
and again the week after that,
and that all the time
his family is safe and in well being
without his personal protection
constitutes a whole new era of evolution—
the first really "new"
since the beginning of the spoken word.
In fact, out of the understanding
innate in the spoken word
was industrialization wrought
after millenniums
of seemingly whichariess spade work.
[The Unfinished Epic of Industrialization]

Tension and Compression are complementary functions of structure. Therefore as functions they only co-exist. When pulling a tensional rope its girth contracts in compression. When we load a column in compression its girth tends to expand in tension. When we investigate tension and compression, we find that compression members, as you all know as architects, have very limited lengths in relation to their cross section sections. They get too long and too slender and will readily break. Tension members, when you pull them, tend to pull, approximately (almost but never entirely), straight instead of trying to curve more and more as do too thin compression members. Therefore as we load a column in compression the contraction of the tension members in their girth, when axially loaded columns. The contraction of the tension members in their girth makes it even stronger. There a no limit ratio of cross section to length in tensional members of structural systems. There is a fundamental limit ratio in compression. Therefore when nature has very large tasks to do, such as cohering the solar system or the universe she arranges her structural systems both in the microcosm and macrocosm in the following manner. Nature has compression operating in little remotely positioned islands, as high energy concentrations, such as the earth and other planets, in the macrocosm; or as islanded electrons, or protons or other atomic nuclear components in the microcosm while cohering the whole universal system, both macro and micro, of mutually remote, compressional, and off non-simultaneous, islands by open prehensive tension, —compression islands in a non-simultaneous universe of tension. The Universe is a tensegrity.
[WDSD Document 2]

# ABONOS VERDES

HAY PLANTAS QUE MEJORAN MÁS RIEGA LA TIERRA QUE FORMAN NITRÓGENO DEL AIRE Y LO PONEN EN EL SUELO.

LAS LEGUMBRES SE CONOCEN MUY FÁCIL PORQUE ECHAN UNA VAINA EN LA QUE VIENEN LAS SEMILLAS.

ESAS TODAS ESTAS PLANTAS SON LEGUMBRES, COMO EL FRÍJOL Y EL CHÍCHARO.

SON PLANTAS MARAVILLOSAS CON ESAS RAÍCES QUE SE METEN MUY HONDO MEJORAN LA TIERRA Y HACEN QUE EL AGUA ENTRE MEJOR.

TOMAN MUCHO NITRÓGENO. UNA HECTÁREA DE LEGUMBRE TOMA DEL AIRE LA MISMA CANTIDAD DE NITRÓGENO QUE HAY EN VEINTE TONELADAS DE ESTIÉRCOL.

EL AIRE QUE RESPIRAMOS TIENE MUCHÍSIMO NITRÓGENO Y ESTAS PLANTAS TOMAN DE AHÍ ES QUE NACEN TAN... Y HASTA PONEN UN POCO EN EL SUELO.

POR ESO ES QUE SE LLAMAN ABONOS VERDES.

## VILLAGE TECHNOLOGY

Most so-called appropriate technologies have been inherited from the past. This community solar still, which provides fresh-

water from seawater, was built in 1969 in Haiti (16); it is virtually indistinguishable from the first such still, built by Carlos Williams in Chile in 1872.

17

Bio-gas plants were developed in Germany during World War II and introduced into India, and subsequently throughout Asia, in the 1950s. A large plant (note the human figure at left) outside Manila illustrates an important aspect of bio-gas technology (17). Considerably more success has been had with large plants than with extremely small bio-gas digesters. This plant converts the manure of 7,500 pigs into gas which runs a pump, generator, and four freezers, as well as providing fuel for lighting and cooking.

over-all modernization strategy. The intermediate technology chosen should not be one that blocks future development; it should not be a goal in itself. When the international development organizations such as the World Bank or the United Nations agencies refer to "appropriate technologies," it is almost always in this context,[8] but when *A Handbook on Appropriate Technology* (Ottawa, 1976) calls for "an alternative approach to development" it obviously implies something more.

What are some of the elements of this "alternative approach"? There is an assumption that in all cases decentralization is desirable. There is a tendency to ignore the technological limitations (not all processes can be miniaturized or decentralized) and to ignore the facts of geography. A small country (such as Cuba), a country composed of islands with navigational contacts (such as the Philippines), or a country with good transportation networks (such as Britain) would be foolish to decentralize production units when goods can be readily transported throughout the region. This is obviously *not* the case for a very large country with poor transportation links (such as China or India).

An overemphasis on simplicity and labor-intensiveness

[8] "The use of appropriate technologies consequently demands a recognition on the part of technology users in developing countries that in order to improve the lot of the vast majority of people they must, *at least in the short-run* [my emphasis], accept standards of service and levels of 'modernity' lower than those that might be found in more developed countries." *Appropriate Technology in World Bank Activities,* July 19, 1976 (unpublished report of the World Bank, Washington, D.C.). Clearly the World Bank has not been taken in by the semantic obfuscation of "appropriate" versus "intermediate." It recognizes that appropriate technology is an intermediate stage; hence, almost by definition, it represents a lower (intermediate) standard of modernization. (But not always; see Chapter 5.)

also sometimes ignores the fact that most less developed countries (particularly in Africa) lack not only capital, but also a properly educated labor force. Wassily Leontief, an American economist, has observed, in *Theories and Theorizing: In Economics* (New York, 1966) that whereas mechanization of nineteenth-century processes required a proportionately large capital expenditure, automation of most contemporary industries requires only a small additional investment (6 per cent or less). Leontief describes a situation in which automation could be used by the less developed country as an intermediate technology, which would result in an anomalous (but not necessarily undesirable) situation in which "economic efficiency may, at least temporarily, run far ahead of progress toward social maturity and stability." This process can be observed today in South Korea, Singapore, and Hong Kong.

According to *A Handbook on Appropriate Technology,* "The Appropriate Technology approach recognizes that different countries and communities have differing cultural backgrounds, priorities, and motivational values, into which a technology must be integrated." Of course, taken at face value, a statement like this is a truism. Even the transnational corporations are sensitive to "different cultural backgrounds, priorities, and motivational values,"[9] but few would claim they espouse a cause,

---

[9] A trivial but revealing example is the report that the Pepsi-Cola company seriously considered abandoning its popular "Feeling Free" slogan in some of its international operations, since it was felt that in certain less developed countries (or in the Soviet Union) this might be interpreted as an endorsement of urban terrorism or liberation movements! On the other hand, Frances Moore Lappé et al. in *Food First* (Boston, 1977) describe how in Brazil the slogan "Pepsi Generation" was changed to "Pepsi Revolution" precisely to foster "protest through consumption."

which, according to the *Handbook,* "does not take for
granted, or impose, the values of industrialized societies,
such as the motivation toward material advancement."
This incredible assertion ignores the fact that development is precisely "material advancement" and that this is
the stated goal of virtually every country in the world
today, irrespective of race, religion, or politics. Such a
neurotic attitude toward industrialization is often translated into a romantic view of "indigenous" technology,
which tends, in practice, to be a preference for "going native" (an unfortunate colonial phrase but doubly accurate
in this context) to the neglect of even common sense. This
emphasis on local particularities may well serve as a
block to the kind of change that modernization requires.

A final word needs to be said concerning environmental
considerations. Since the publication of Schumacher's
*Small Is Beautiful* in 1973, environmentalists have assumed that small technologies will result in less disruption
of the physical environment. The limited accuracy of this
statement has already been pointed out, but more damag-
ing is the fact that it ignores the stated priorities of the
less developed countries: the fact is that they are less concerned about environmental pollution than the advanced
countries. The Founex Report, which was presented to
the UN Stockholm Conference on the Human Environment in 1972, stated the position of the so-called Third
World countries in which "the major environmental problems are of a different kind: they are predominantly problems that reflect the poverty and very lack of development
of their societies." Statements such as this make it quite
clear where the less developed countries stand. There is
always a limit to the price people, or states, are willing to
pay to avoid risks. This limit is proportionately higher in
the richer countries than in the poorer countries. Pollution

abatement is not high on the list of priorities of most of the less developed countries. It may be argued that it should be, but if one is concerned with "differing motivational values," one must put a low value on environmental impact when evaluating a technology for most less developed countries. The fact that some environmentalists do not do this not only illustrates their intellectual dishonesty, but risks seriously penalizing the group that adopts the ecological (*sic*) technology, since it is paying for something that it neither wants nor, many would argue, needs.

## LET THEM DRINK COKE

There is little evidence that "undoubtedly appropriate technology . . . has its modern roots in the developing world," as one publication claims.[10] The vast majority of AT advocates and practitioners are to be found in the West; what is more, many of the groups in the less developed countries are dependent on foreign financial support.[11] There is an extensive, long-established network

[10] R. J. Congdon, ed., *Introduction to Appropriate Technology: Toward a Simpler Life-style* (Emmaus, Pa., 1977).

[11] The *Introduction to Appropriate Technology* deals specifically with technologies for less developed countries. It lists thirty-five appropriate technology groups. *One* of these is located outside Europe or North America; twenty-five are to be found in the United States. A later publication, *A Handbook on Appropriate Technology*, referred to above, lists eighty-four groups; half are from the industrialized, advanced countries. In spite of this, the *Handbook* claims that "Appropriate Technology will have the advantage of reducing a developing area's economic and cultural dependence on industrialized nations and their modes of operation." I once spoke with the local director of an AT center in South Asia who told me that if the subsidy he was receiving from an international church group stopped, he would have to close down.

of programs in less developed countries to encourage small industries, but these, on the whole, share little of the "small is beautiful" ideology.[12] Finally, the pronouncements of the less developed countries themselves, such as those of the "Group of 77," which at a series of United Nations Conferences on Trade and Development (UNCTAD), has called for an increase in the transfer of *advanced* technology from the more developed to the less developed nations.

The risk for a less developed country of using intermediate technology is when the latter is a well-intentioned but ill-informed, imposition *from outside,* an idea of "the rich about the poor," and instead of being a stepping stone to further development there is a danger that it will become a millstone around the necks of the poor in the less developed countries.[13]

What are the reasons for making such a pessimistic prediction? All right, the reader may say, perhaps some of the assumptions of Western AT groups *are* inappropriate and do not always reflect the realities, or the desires, of the less developed countries. Perhaps the emphasis on "ecology" is misplaced, given the problems of the poor. But surely each country will simply modify and redirect AT to suit its particular purposes, such as China did after

[12] The principal aim of most small industry development institutes in the less developed countries is not only to foster small industry but to encourage and assist small industry to grow into *large* industry.

[13] "Certain subjects, like poverty and intermediate technology, keep the experts busy. They are harassed by international seminars and conferences and foundation fellowships. The rich countries pay; they dictate the guiding ideas, which are the ideas of the rich about the poor; ideas sometimes about what is good for the poor, and sometimes no more than expressions of alarm" V. S. Naipaul, *India: A Wounded Civilization* (New York, 1977). ("Harassed" is good!)

the Great Leap Forward or India after the Community Development movement (1950–65)? The crucial difference is that these two movements were both indigenous in origin—this did not make them more correct, but it did tend to make them more correctable.

The theory of Appropriate Technology, on the other hand, often comes to the least developed country from the most advanced countries of North America and Europe.[14] Herein lies one part of the imposition. A well-thumbed copy of The Whole Earth Catalog (Stewart Brand, ed., Millerton, N.Y., 1968), no less than a poster of the former model and TV star Farrah Fawcett, reflects the glamour that is attached to all things Western. There is no escaping this. Transnational companies seduce the uneducated peasant into buying expensive soft drinks with money that reduces an already meager diet—mothers feed children Coca-Cola because they believe the manufacturer's slogans. The effect of glossy publications, not on peasants but on the upper-class ecology-conscious young, is not so different. They also believe slogans. Their icons—windmills, composting toilets, solar stills—are constructed (often incorrectly) and proudly shown to visitors. As much as anything else, they are a sign of a temporary escape from the provincial environment of what the Trinidad-born Indian writer V. S. Naipaul has called "the overcrowded barracoon."

Some of the support for the transfer of these ideas from the West comes from those disaffected with modernization, who see this as an opportunity to promote their

[14] AT publications from advanced countries, especially Britain and the United States, are high-quality, expensively produced consumer items. It is hardly surprising that they have a greater impact in the less developed country than locally produced pamphlets, usually amateurishly printed on a shoestring budget.

ideas in other countries.[15] Support has also come from the personnel of various United Nations agencies. It is not clear whether this is out of political, professional, or personal interest. The United Nations tends to be staffed by people who are the ultimate expatriates but who see themselves as the agents of the underdeveloped world, and most of these interested parties are also active supporters of foreign aid, one of the vehicles for technology transfer to the less developed countries.


## FOREIGN AID

There are three basic positions taken by critics of foreign aid. The first is that the richer countries do not give enough aid and that they should give more (not surprisingly, this is the position of many of the less developed countries who also call for greater equality in the distribution of resources). The second position, taken by conservative critics in the advanced countries, is that there should be less, not more, foreign aid. According to this argument, foreign aid is really welfare which retards, rather than encourages, development. The third position, taken by some liberal writers, is critical of foreign aid itself, on the grounds that foreign aid is usually politically motivated and economically benefits the donor as much as the receiver. A discussion of these three positions would be lengthy,[16] but a few words need to be said on the sub-

[15] The promotion of bankrupt ideas in less developed countries is nothing new. For instance, architects from America and Europe have for a long time, and with little success, attempted to promote new geometric forms under the guise of emergency housing. This is described by Ian Davis in *Shelter After Disaster* (Oxford, Eng., 1978).

[16] The call for more aid is regularly made by most liberal writers. The call for less aid has been made by P. T. Bauer in *Dis-*

ject of bilateral aid, for this is how many AT projects have been supported.

There have been two basic types of foreign aid: *multilateral* and *bilateral*. Multilateral aid is administered by organizations such as the World Bank or the various agencies of the United Nations—there is no direct contact between the donor countries who support these organizations and the recipient of a loan or grant. Bilateral aid is given by one (more developed) country to another (less developed) country. Bilateral aid is usually administered by government ministries or departments such as the Canadian International Development Agency, the Agency for International Development (United States), the Ministry for Overseas Development (Britain), the Ministry for Development Cooperation (Netherlands), and so on. Although foreign aid conjures up images of grain and powdered milk, this is not always the case. For instance, during the 1960s *one quarter* of all American foreign aid went to one country—South Vietnam—and this largely in the form of armaments. Today, one quarter of all U.S. aid goes to two countries—Egypt and Israel—and again largely in the form of military hardware.

The view that all foreign aid is in the form of gifts is likewise mistaken; most aid is actually in the form of loans, some interest-free, some low-interest, and others with normal interest charges. Moreover, the vast majority of these loans are not in cash; they are credits on future sales. This is called "procurement tying"—that is, the recipient country receives a loan or gift which can be spent only on goods and services from the donor country. In

_____

*sent on Development* (London, 1971). Criticism of current foreign aid practices can be found in Michael Harrington's *The Vast Majority* (New York, 1977). Gunnar Myrdal has argued in a number of his books for more multilateral and less bilateral aid.

some cases aid is "double-tied"—that is, it is related to a
particular project and must be spent on specified hardware
and services. Given this situation, it is hard to dispute the
claim of critics that the beneficiary of foreign aid is often
the donor country. One finds Chevrolets in the Philippines, Renaults in Senegal, Land-Rovers in Kenya, and
Moskviches in Cuba—a reflection of colonial history and
political alignments.

Virtually all bilateral aid is procurement-tied, which
has several economic implications for the recipient country.[17] Since it has little choice in where—and often
how—to spend the money, there is no opportunity to
"shop around." The best and cheapest technology may be
Japanese, but if the aid is from France, then the technology must be French. It is difficult to see how AT, as a
constituent of foreign aid, is going to steer clear of such a
pitfall. There is a great danger that intermediate technology will simply become part of the "aid package," in
which case there is no guarantee that appropriate criteria
can and will be adhered to. This is not because of ill will
or any conspiracy on the part of the more developed
countries, but rather the result of the nature of bilateral
aid, which, in practice at least, is nationalistic and protective of self-interest.[18]

[17] Almost all American aid is procurement-tied, as is the majority of British and French aid. Dutch and Canadian aid is more
evenhanded, though tying is common. Sweden alone imposes virtually no restrictions on its foreign aid.

[18] One should not imagine that self-interested aid is restricted to
capitalist societies. In 1976 the People's Republic of China com-
pleted a 1,100-mile railway from Zambia to the Tanzanian coast.
*All* the engineering hardware was imported from China. Within
two years 30 per cent of the rolling stock was out of commission
because of maintenance problems and lack of spare parts. See
*Time,* November 6, 1978.

Even if AT is part of a bilateral program that is not procurement-tied, the problems are formidable. In 1976 the United States Congress formed a new organization called AT International to support intermediate technology activities in the less developed countries. It appears that AT International is making efforts to promote appropriate technologies without simply promoting American manufacture and export of such technologies and is working directly with entrepreneurs in less developed countries and planning to "focus on small groups in the hope that the sparks of innovation that are ignited will fire the imagination of larger groups until whole societies are involved."[19] This optimistic statement begs the question of whether the political and economic goals of the host state coincide with those of AT International. Even if "the host government does not oppose in principle the implementation of the project," this is hardly the basis for embarking on such ambitious plans. Is it in fact possible to promote "self-sufficiency," "local initiative," or "local control" from outside, through bilateral foreign aid? I doubt it. As an early president of the World Bank, Eugene R. Black, put it, "But even at best, there is always the risk that political influences may misdirect [bilateral] development aid, since they may bring in considerations that are irrelevant to the real needs."

This chapter may have struck the reader as unnecessarily critical, but I cannot minimize the ill effects that a stubborn and willful application of preconceived ideas about what is an "appropriate" technology could have on the less developed countries. These countries, and particularly the poor of these countries, do not have the re-

[19] "AT International: An Overview," March 6, 1978, unpublished report. On the other hand, the board of directors of AT International consists solely of American citizens.

sources for experimentation or for error. There are already a number of examples of tragic, though often well-meaning, impositions made in the name of progress or modernization. It would be sad indeed if the idea of Appropriate Technology, which has something to offer, were to become another such misapplied paregoric.

It does not have to be so.

First, advocates of intermediate technology should look for alternatives to foreign aid as vehicles for technology transfer. If people are to choose technologies freely, then they must be allowed to make the choice themselves, by themselves. The imbalance of foreign aid, particularly bilateral aid, makes such a free choice difficult. There is nothing wrong with European or American research groups developing intermediate technologies for use in less developed countries, provided that these technologies are chosen for use by the people in those countries and not by outside national or international aid agencies.

Secondly, it is necessary to reassess some of the assumptions of the Appropriate Technology movement about the nature of development and about the nature of technology. The need for a technology, or technologies, scaled to the resources and needs of the world's poor is undoubted, but it is not necessarily useful to describe this as a different *type* of technology. Likewise, it is inaccurate to imagine such technologies as being the basis for a totally different type of development. There is little indication that such a development is possible; there is even less evidence that most less developed countries find it desirable.

These are hard lessons. Though they seem to call into question some of the most widely held views of the AT movement, they by no means refute the basic belief that it is the small technologies, not the big, which may tip the balance toward real development in very many countries. The paramount example of "the taming power of the small" has been the often cited, but not always well understood, technological development in the People's Republic of China.

*Chapter 3*

## ... OR STEPPING STONE

"Walking on Two Legs"

—*Industrial slogan from the People's Republic of China*

It is now necessary to make an extended detour, eastward, to examine the role that small-scale industries have played in the People's Republic of China. The reader of publications on intermediate technology cannot help but be struck by the number of references made to China: the Chinese ride bicycles (true); the Chinese substitute labor for machinery (partly true); there is no profit motive in China (largely untrue); the Chinese use small rather than large technologies for ideological reasons (untrue). There is no doubt that the myth and reality have had an important influence—partly as inspiration, and partly as proof— the rare case of intermediate technology being applied on a national scale. The inspirational aspects do not need clarification; the "proof" does. To paraphrase Texas Guinan, the American burlesque star of the 1930s, can one billion Chinese be wrong? The answer is yes and no. The Chinese approach definitely proves the value of intermediate technology, but it would be a mistake to imagine that China is a "test case" for antimodernization. In China the "small" approach has been adopted chiefly in a pragmatic way, hand in hand with industrialization, not as an end in itself. In China, small is not beautiful—it is only necessary.[1]

One can attribute the Sinophilia of certain neo-Utopian writers to the current fashion of China-gazing. This tendency is especially pronounced among those of the left and is usually characterized by an uninformed and uncritical admiration for all things Chinese, which are invariably portrayed as antidotes to all the perceived ailments of modernization. This in spite of the fact that modernization in China has scarcely begun; as Simon Leys, a Belgian art historian who lived in the People's Republic of China for ten years, has caustically remarked, "One might as well praise an amputee because his feet aren't dirty."[2]

Comparisons between China and the most advanced countries are, at least for the moment, fallacious and need not be taken seriously. The same cannot be said for comparisons between China and other less developed countries; the Chinese approach to development in this context demands, and has received, much serious study. However, the difficulties of acquiring and interpreting information on China are formidable. An American delegation may go

[1] Chinese pragmatism on this point is illustrated by a remark made by First Deputy Premier Teng Hsiao-p'ing concerning the backwardness of his country: "If you have an ugly face it is no use pretending you are handsome" (Time, November 6, 1978).

[2] "Western ideologues now use Maoist China just as the eighteenth-century philosophers used Confucian China: as a myth, an abstract ideal projection, a utopia which allows them to denounce everything that is bad in the West without taking the trouble to think for themselves. We stifle in the miasma of industrial civilization, our cities rot, our roads are blocked by the insane proliferation of cars, et cetera. So they hurry to celebrate the People's Republic, where pollution, delinquency, and traffic problems are non-existent. One might as well praise an amputee because his feet aren't dirty." Simon Leys, Chinese Shadows (New York, 1977). "Simon Leys" is the pen name of Pierre Ryckmans.

specifically to study small rural industries, yet any conclusions are seriously qualified, for it may have visited only,
say, 60 establishments out of more than 200,000. For a
country as huge and spread out as China, the reliability of
statistics must also be questioned, by the Chinese themselves no less than by outsiders. Scholars trying to inter-
pret data perform feats of Holmesian deduction to un-
ravel reality from propaganda. All this is not necessarily
the result of deviousness or inscrutability on the part of the
Chinese (though at certain times this has been the case);
it is a problem common to the whole developing world
where limited resources, carelessness, and sometimes national pride conspire to compromise seriously the scientific value of many statistics.

Beyond the statistical there lies the even more distant
physical reality. "But it works in China" is a frequent
clumsy rejoinder to expressed doubts about a particular
technology. Yes, but *how* does it work? Everyone
"knows" that there are thousands (who counted them?)
of composting plants in China; the reality of what it must
be like to use them tends to be obscure.[3] One is told that

[3] There is one unique description of Chinese composting from
the user's point of view: "I felt like vomiting. Hundreds of grunt-
ing, snorting [pigs], massive and black, were struggling to get at
the potato peels, wild vegetables and miscellaneous garbage that a
prisoner was heaving into their troughs. It was hard to figure
which smelled the worst, the garbage, the pigs themselves or the
excrement. The pig yard was made of brick, like a patio, so that
none of the fertilizer would be lost, and their living quarters, rows
of slant-roofed pens, were carefully freshened up each day by
shovels of fine, dry, sandy earth. Our job was to shovel the sand
from the pens after the pigs had fouled it, toss the wet mess into a
ditch over on the side and then add straw to the mixture. The pig
excrement and urine quickly fermented with the straw to make a
horrible, rich, black mess that was high-grade fertilizer. We would
then scoop the muck out, pile it in mounds and repeat the process.
There are many ways of building socialism." Bao Ruo-Wang

the Chinese are making a New Man; the facts rarely confirm this.[4] Perhaps I am belaboring all this, but the fact that rational discussion of China has become increasingly difficult needs to be pointed out. There are important lessons to be gained from the Chinese experience *if* one approaches the issues without preconceptions. I am aware of the pitfalls of generalizing in so short a space about a country and a people that virtually encompass a continent, and I do so with the full knowledge that much of this data is probably incomplete and that some of the interpretations will have to be amended in the future as new facts come to light. The reader has been warned.

## SMALL INDUSTRY IN CHINA

A singular aspect of Chinese industrialization has attracted a great amount of attention: the growing importance of rural small-scale industry as a tool for development.[5] "Small industry" is a rather loose label that describes factory-workshops that employ anywhere from a dozen to 500 workers. Though small size is a common denominator, their chief common characteristic is their management. Unlike large Chinese industry, which, following the Soviet model, is completely centralized, these

---

[J. Pasqualini] and Rudolph Chelminski, *Prisoner of Mao* (New York, 1973). Pasqualini, the son of a Corsican father and a Chinese mother, spent seven years (1957–64) in Communist Chinese labor camps.

[4] The Chinese have remained, for instance, inveterate cigarette smokers. This is heartening, though unhealthy; the faultless New Man, were he to exist, would probably be insufferable.

[5] This importance is not inconsiderable: "By 1966, two thirds of the gross value of agricultural machinery production came from local medium and small plants." Carl Riskin, "Small Industry and the Chinese Model of Development," *China Quarterly*, No. 46, April/June 1971.

industries are managed at the provincial, or lower, level.
From this important (particularly in the Chinese context)
fact result other attributes. The small industries "serve"
agriculture; that is, they manufacture consumer and pro-
ducer goods for their immediate, local market. The small
industries are self-supporting—they receive only occa-
sional inputs from the central industries and must make
do on their own, which leads to relatively labor-intensive
technologies and use of local materials. The small industries must also use local manpower, which in the rural
areas means primarily off-season agricultural workers.

The small industry strategy has had many obvious advantages, which will be fully discussed later. The most important of these is that it has allowed a very large country
with a very poor transportation network and widely
diffused resources—but with a unique human resource, the
hard-working and ingenious Chinese peasant—to begin on
a path to industrialization.[6]

It should be emphasized that the small industries program, in spite of the attention that has been given to its
ideological aspects, is a profoundly common-sense reaction to the reality of Chinese circumstances. It is likely
that *any* Chinese state, whatever its political goals, would
end up with some kind of small, decentralized, rural in-

[6] The impression is sometimes given that, for China, industrialization is just around the corner. Nothing could be further
from the truth. China today has one sixth the road mileage, one
third the steel production, and one quarter the electrical generating
capacity of the United States in *1940*. Nevertheless, the first opinion
poll ever conducted in the People's Republic indicated that 80 per
cent of persons questioned believed that massive modernization
could be achieved by the end of the twentieth century (Agence
France Presse, January 18, 1979). This indicates that in China, as
in most less developed countries, there is a wide consensus about
the desirability of modernizing in spite of the prodigious obstacles
that stand in the way.

dustry approach.[7] As it turned out, the People's Republic of China arrived at this policy after a number of false starts.

The First Five-Year Plan (1953–57) was a doctrinaire, Soviet-inspired and Soviet-supported effort to achieve massive industrialization. Until that time, Chinese efforts at industrialization, which had begun in the 1920s, had been small. Although there are indications that the First Five-Year Plan was to accommodate small rural industry and agricultural development, in practice the major emphasis was on heavy industry. The rural industries, which were supposed to support agricultural development, were generally neglected and, as a result, agricultural production stagnated. By 1957 it was becoming clear to the ruling elite that a mistake had been made; the traditional base of the Chinese economy could not be ignored. At that point a choice had to be made—either heavy industry (whose growth rate, despite the effort of the Plan, remained slow) had to be diverted to support agriculture or much greater emphasis would have to be put on raising agricultural production at the local level.

The Chinese chose the latter, more or less. Mao Tsetung's Great Leap Forward (1958–60) was a combina-

[7] Two professors, one an Englishman and the other an American, were able to write in 1944, *five years before the Communists came to power* and fourteen years before the Great Leap Forward: "The future relationship, however, of agriculture and manufacturing industry [in China] is a matter of great interest. Development of electric power and improvement of transport would render possible diffusion of *small-scale industries* [my emphasis] better suited to Chinese traditions and genius than the large-scale factory system, and, if accompanied by the growth of *co-operative agencies* [my emphasis], may greatly improve the conditions of the countryside." K. S. Latourette and P. M. Roxby, "China, III: Production, Commerce and Communications," Encyclopaedia Britannica (Chicago, 1949).

tion of a renewed emphasis on rural development and a rabid antimodernization (mostly anti-Soviet) characterized by a very un-Chinese loss of pragmatism.[8] There were mistakes made in this rushed attempt to "ruralize" industry. According to Carl Riskin, an American economist at Columbia University, industries whose technological spectrum did not include small-scale and labor-intensive options were chosen for local development, so that there were a good many resources wasted in the production of goods of inferior quality. For instance, millions of Chinese were encouraged to build backyard iron and steel plants, but the material produced was such low quality as to be virtually useless. The resource base for rural industrialization was likewise problematic. There were hardly any "surpluses," either of labor or materials, since the population as a whole was living on the margin of subsistence. As a result, the main resource base for rural industrialization turned out to be the traditional handicraft industries, which in 1956 included over 5 million workshops.[9] The Great Leap Forward simply engulfed this sector, with predictable results as regards disruption of production. Likewise, many agricultural workers were shifted to communal and provincial industries.[10] The results of the Great Leap Forward on the Chinese economy were

[8] I use the terms "Great Leap Forward" and "Cultural Revolution" with misgivings and only because they are widely recognized. The former would more accurately be described as a "great leap *backward*" and the latter was less revolutionary than anarchistic.

[9] The role of small craftsmen remained important, according to Riskin: in 1962, more than 80 per cent of all small farm tools were being manufactured by handicraftsmen.

[10] Some of these shifts were enormous. In one province, between 1958 and 1960, 2 *million* workers were reportedly shifted from agricultural production to iron and steel factories in communes and factories.

devastating. According to Simon Leys, "Not only did the movement fail to achieve the exhilarating aims it had set itself, but the entire Chinese economy was plunged into chaos when the construction effort met paralysis and breakdown."[11]

The next period in Chinese development, immediately preceding the Cultural Revolution in 1966, represents a positive stage as regards the growth of small industries. The excesses of the Great Leap Forward were corrected; agricultural workers were returned to the farms and the local handicraft industries were allowed to return to their original "small" state. A greater rationalism was introduced into choice of industries for labor-intensive production. Without returning to the inappropriateness of the First Five-Year Plan, there was a veering away from the romantic irrationality of the Great Leap Forward, while at the same time maintaining an emphasis on agricultural, as opposed to industrial, development.[12]

During the turbulent period of the Cultural Revolution of the late 1960s and the political in-fighting that followed, up to and immediately after Mao's death in September 1976, the basic policy with regard to small rural industry did not change. Whatever the immediate economic effects of this neurotic decade of revolution, it is

[11] Leys, *The Chairman's New Clothes: Mao and the Cultural Revolution* (Paris, 1971). The results were also lasting: "The Great Leap and subsequent depression cost Red China at least several years, and perhaps as many as six or seven, in overall economic growth and industrial production" (B. M. Richman, *Industrial Society in Communism* [New York, 1969]). Nevertheless, Richman is forced to conclude that "few developing countries have done as well as China in growth and development since 1950."

[12] Incidentally, during the period 1959–68 Mao, though still exerting influence, had been replaced as head of state by Liu Shao-ch'i.

likely that the long-term influence on mass education and development will be significant. The foundation for the industrialization of China's countryside was laid during this period, and, whatever the optimistic prognostications of Mao's successors, three quarters of China's population still live outside the cities. If progress will be made, it must be made in the villages or it is unlikely to be made at all.

To summarize, small industries play an important role in Chinese development and contribute significantly to industrial production. It is impossible to say whether this approach is for China the "best" one; the growth rate for Chinese industry *and* agriculture has been very low.[13] On the other hand, given the circumstances under which China is beginning to modernize, it is perhaps the best that could have been hoped for. Small industries have several distinct advantages for the Chinese: decentralization has overcome the problem of a very limited transportation and marketing system. It also saves time and resources since plants can be put into operation more quickly, and maintenance and repair downtime is reduced since it is done locally. The small plants can make use of local resources, even if the latter are not abundant enough to warrant exploitation by heavy industry. The products of small industry, since they are sold locally, can be suited to special market needs; China is, after all, a varied country of different climates, topographies, and cultures. Finally but probably most importantly, rural small industries are part of an over-all strategy for narrowing the gap between

[13] The growth for industrial production between 1957 and 1970 was 4–5 per cent per year; the growth rate for industry and agriculture combined was only 3–3.5 per cent for the same period. "Quarterly Chronicle and Documentation," *China Quarterly*, No. 46, April/June 1971.

the city and the countryside, an endemic problem of most less developed countries.

## WALKING ON TWO LEGS

The most interesting lessons to be learned from the Chinese experience are the particular way in which intermediate technology has been used and its relationship to heavy industry and over-all modernization. Paradoxically perhaps, there is little that is doctrinaire about the Chinese approach, which has been described as "walking on two legs."

First, the small industries program in China is not an across-the-board approach. After the Great Leap Forward, the Chinese made two important discoveries: only certain industries were suited to small-scale production and these small industries ought to produce primarily for the agricultural sector. The small industries fell into four main categories: cement plants, agricultural machinery fabrication, fertilizer production, and a general category which included capital projects such as irrigation, roads, and building. All of these lent themselves to a certain level of miniaturization. Conversely, certain industries which had originally been small, such as hydroelectric power plants and steel plants, were enlarged to achieve a minimum viable size. The gearing of small industry to serve, and improve, agricultural production is an important point. The Chinese recognized that, as in most less developed countries, agriculture was, and would be for a long time, the base for economic development. Thus the small industries were not (as during the Great Leap Forward) an attempt to miniaturize heavy industry, but rather an attempt to reinforce agriculture.

This leads to a second point. There are no indications

that the small industries program was an attempt to create employment; in the most rural regions the maximum proportion of the labor force employed in small industries was 10 per cent, in others much less. The only exceptions were large capital projects, such as irrigation canals, which did use a great deal of labor, but only for limited periods of time. These capital-creating projects were usually planned to take advantage of slack periods when agricultural labor was free. The main purpose of the small industries was to provide the means—fertilizer, farm machinery, cement—whereby the labor productivity of agriculture could be significantly increased.

This approach quite obviously stressed *productivity,* hence the Chinese predilection for mechanization in both small industry *and* agriculture. The rationale for the mechanization of agriculture, which is attributed to Mao, goes like this. If agricultural productivity is raised through improved techniques (irrigation, fertilizer, high-yield seeds—i.e., the "green revolution" technique) but with continued high labor inputs, a continuing low level of labor productivity is implied. The increase in income to the individual peasant would be marginal, and incentives to increase production would accordingly be minimal. On the other hand, if mechanization is introduced, yields go up much more, the labor input drops, and labor productivity per capita rises. This is a refutation of Gunnar Myrdal's proposal that in developing countries *only* agriculture can absorb more labor; if twice as many farmers produce twice as much, no progress—from the individual farmer's point of view—has been made. It is precisely the individual farmer whom the Chinese had in mind, for they realized that for the standard of living of the peasant to rise, his individual productivity must rise, not by fractions but by factors of five, ten, or more.[14] Obviously, mechanization displaces labor, and this must be absorbed by a diversification of the rural economy. Mechanization is likewise an important part of the small industries. The Chinese have stated explicitly that they are interested in the output per worker and are not concerned about displacing labor.[15]

It might be worth mentioning parenthetically that the importance of productivity was closely linked to the incentives present in the Chinese system. There was emphasis on self-support at every level (team, brigade, commune, and province) and capital funds tended to be generated at that level. Thus there was a built-in feature, similar in some respects to the profit motive in capitalist societies, to be concerned with efficiency and productivity. Ever since the Great Leap Forward the amount of profit that the central government has "allowed" the small industries to keep has steadily increased. The taxation system is extremely regressive—profits above a certain

[14] The farmer's view on mechanization is expressed by a Chinese peasant: *"The new agricultural implements make farmwork easier* [my emphasis]. We are hoping for tractors. If you come back in ten years' time we shall be working all the land down here in the valley by tractor. The same thing will happen with manuring . . . We are going to drive our millstones with electricity instead of by hand or donkey. Life will be much better then." Jan Myrdal, *Report from a Chinese Village*, trans. Maurice Michael (New York, 1965).

[15] "The Chinese maintain that mechanization only frees people for more important tasks and that more and more work will remain to be done. In the short run, and at the presently low levels of mechanization in most places, they are undoubtedly correct in asserting that mechanization *creates* a demand for rather than *reduces* the demand for labor. In the long run, industry may well have begun to develop rapidly enough to absorb any displaced labor." The American Rural Small-Scale Industry Delegation, eds., *Rural Small-Scale Industry in the People's Republic of China* (Berkeley, Calif., 1977).

amount are virtually tax-free—entrepreneurship is encouraged (as it would be in any economy), and efficiency and productivity are sought for. These incentives are carried further in the collective industries, where a system of "work points" distributes the profits of work teams among its members. The fact that the village industries have, for all practical purposes, a protected market and that the work-point system is in effect a local money supply prompted one commentator to describe this, quite accurately, as "village Keynesianism."[16]

It should be obvious by now that the Chinese approach to using small industries is not a rejection of modernization. There was a moment during the Cultural Revolution when *t'u* (native, indigenous) was promoted in favor of *yang* (modern, developed). This now seems to have given way to an idea of development from *hsiao-t'u* (small native) to *hsiao-yang* (small modern). This differentiation between modern and indigenous is critical. At the same time, the emphasis on small industries in China in no way comprises a turning away from large-scale industrialization. There has never been a point since 1949 when China has abandoned the idea of large, centralized heavy industry. Even during and after the Great Leap Forward, when so much industry was decentralized and removed beyond the authority of the state, heavy industry was firmly retained under state control. This two-tier system is best represented in the Ta-ch'ing oil fields (a "model" project), where the petroleum industry is controlled by the state, but the self-sufficiency of the industrial city surrounding the fields is provided for by locally controlled small industry and agriculture. This is what is described as "walking on two legs": a reliance on

[16] *The Economist*, December 13, 1977.

small-scale, labor-intensive methods, as well as on large-scale, capital-intensive techniques. Significantly, the former is seen to be temporary and leads to the latter.[17]

This idea of small-scale technology as a stepping stone to ultimate large-scale industrialization is very similar to the theory of intermediate technology, though in practice it puts more emphasis on mechanization and productivity than the latter does. It also recognizes that there are intrinsic limits to the small-scale approach. There is a limit to how much productivity can be increased with continued labor inputs—at some point mechanization must be introduced. Small tractors increase productivity, but only until the next threshold is reached. The technology of small cement plants produces a rather low-strength cement, quite adequate for small rural works, but when larger projects (e.g., grain elevators) are required, the higher-strength cement needed will require shifting to a different production technology, which may no longer be efficient at the reduced scale.

Since the fall of the so-called Gang of Four in July 1977 there has been a dramatic and significant shift in China's policies toward development, the cumulative effect of which is a *de facto* repudiation of the Cultural Revolution. The stated priorities of the new regime are the "Four Modernizations"—agriculture, industry, national defense, and science and technology—a program

[17] The following appeared in the Peking newspaper *Ta kung pao* as early as May 18, 1959: "It is the direction we follow in development to build large-scale enterprises using modern production methods; but at a time when our capital and technical conditions are not yet adequate, the development of small-scale enterprises using native production methods is the goal of our major efforts for a certain period of time and in given places." Quoted by Carl Riskin in "Small Industry and the Chinese Model of Development."

that had originally been advanced by Chou En-lai. Teng
Hsiao-p'ing, the First Deputy Premier, has been quoted as
saying: "The economy is the goal; politics are only the
way that leads to this goal." The external evidence of
these new policies amounts to a revolution as regards
China's attitude to the West. In 1978 a Sino-Japanese
agreement was signed under which two-way trade would
attain a level of $20 billion over an eight-year period.
Chairman Hua Kuo-feng has called for the completion, by
1985, of 120 main industrial projects, including ten oil
and gas fields, ten iron and steel plants, nine nonferrous
metal complexes, eight coal-mining projects, thirty power
plants, six trunk railways, and five port facilities.[18] This
prodigious growth is to be achieved through the massive
import of foreign technology, especially as regards pet-
rochemicals, iron and steel, aerospace, fertilizers, and
communications. Petroleum extraction is the key industry
in this context, for it is by petroleum sales that China
hopes to pay her bills.

Are the Four Modernizations a repudiation of "walking on two legs"? The interpretation of Chinese politics is
a perilous business; no sooner is one set of pronouncements accepted by Western observers (left or
right) than the seesaw tilts to the other extreme. The
"facts," in either case, tend to be elusive, but it does seem
that intermediate technology in China has a double role to
play. It appears that in many sectors a threshold has been
reached and intermediate solutions will be replaced with
advanced techniques through the import of foreign technology. Only this kind of industrialization is likely to increase productivity and living standards. In such cases, in-

[18] *The Times* (London), September 29, 1978.

termediate technology has served as a stepping stone to modernization.

At the same time, it is likely that in many sectors intermediate technology will continue to be used for some time to come. This is particularly true in the rural areas, where traditional agriculture is so productive that the adoption of more advanced techniques would have less impact than in other parts of the economy. It is also true in education, health services, and various craft industries, where labor-intensive, small-scale techniques are, and presumably will continue to be, used.

China has shown how small industries can play an important role in rural development and how their aggregate effect on national growth is not inconsiderable. On the other hand, it has also shown the fallacy of overemphasizing labor-intensiveness to the detriment of productivity and efficiency. The goal of development is to produce more for each individual worker—only rich countries can afford make-work programs. The Chinese experience has shown that the policy of small-scale industries must be applied pragmatically; it must take into account geography and especially technology. Even if social decentralization is desired, technological decentralization must respect the laws of science; miniaturization of production processes is possible only for a selected number of industries and for certain kinds of products. The small-scale approach in China has, from the beginning, been a tactic, not a strategy. The over-all strategy was toward modernization—intermediate technology was regarded as complementing, not contradicting, this end.

*Chapter 4*

# CALIFORNIA DREAMING

DICK RAYMOND: "By the way, what do you think you'll call it?"
STEWART BRAND: "I dunno—'Whole Earth Catalog' or something."

There are two contemporary positions vis-à-vis modern technology and development which might be described as *evolutionary* and *revolutionary*. The evolutionary position maintains that modern technology is often inappropriate to the needs and resources of most of the less developed countries. It proposes an "appropriate technology" that can serve as a stepping stone to further modernization. This argument is essentially that which was advanced originally as "intermediate technology." The revolutionary case rests on the assumption that modern technology is inappropriate *per se* and that "appropriate technology" represents a substantially different direction for development; this view is an alternative, *not* an intermediate, strategy.

A discussion of these two positions is complicated by the fact that very often people will shift from one to the other, depending on the context and the audience. Thus Schumacher will put forward the evolutionary position when addressing a United Nations group, while when he speaks at a seminar in California or India the revolutionary view will predominate. But one should not jump to the conclusion that politically one position is "left" and the other "right." For instance, the conservative Gandhian position is revolutionary, while the socialist Chinese position seems to be evolutionary; at the same time, the conservative World Bank *is* evolutionary, and the neo-Utopians *do,* on the whole, support the revolutionary view.

The coexistence of the two opposing views is not necessarily hypocritical—it is characteristic of many protest movements, whose recruitment method is often based on sloganeering rather than on intellectual rigor. The black protest movement in the United States linked an evolutionary position (which stressed equal opportunity) to a revolutionary one (which stressed the uniqueness of Afro-American culture). Similar polarities exist in the women's movement (equal rights versus feminism), as well as in many national separatist movements.

The polarization of the AT movement has an important implication. The evolutionary position is concerned, by definition, largely with less developed countries. Since intermediate technology is defined as a stepping stone, it cannot logically be applied in countries which have already "made the step." Some proponents of the evolutionary position go so far as to maintain that the more developed countries (e.g., the United States) have themselves used intermediate technologies to achieve their present state of industrialization. The revolutionary position (alternative technology) can be, and is, applied to the more developed as well as to the less developed countries. "You must change your ways" is the message to the one; "it is never too early to get on the right track" is the advice to the other.

Revolutionary AT as an alternative goal in industrially

advanced societies has not met with a great measure of success. To the extent that it has been accepted, it has been accepted very provisionally and only in specific contexts. The environmentalists, for instance, have supported the soft energy approach, and there is a growing commitment to renewable sources of energy (solar, aeolian, tidal, bio-mass) in countries such as France, Sweden, and the United States. Part of the soft energy argument is for a reduction in nuclear dependency, and recent developments in Austria, Sweden, and the United States have indicated a shift in this direction. But are these signs that a different *technological* strategy is being adopted? I think not. It is true that a different energy strategy is evolving, but this is (necessarily) within the context of advanced industrialization. American transnational companies like Grumman Corporation and Kennecott Copper Corporation have begun manufacturing solar water heaters; a French consortium consisting of Électricité de France, the Commissariat à l'Énergie Atomique, the Centre National d'Études Spatiales, and others is developing solar energy technology; the West German government has built a three-megawatt wind generator on the North Sea coast; the government of Prince Edward Island, in Canada, is considering a thousand-megawatt tidal power installation. My point is not that this disproves the soft energy argument (quite the contrary), but neither does it constitute an endorsement of *any* revolutionary criteria—these devices are neither small, cheap, nor are they particularly "easy to understand." Above all, they do not represent a *volte-face* as regards modernization.

The neo-Utopian left has had less public impact than the environmentalists. The National Center of Appropriate Technology is often pointed to as evidence of the growing acceptance of AT in the United States. The Center is supported by the federal Community Services Administration, which replaced the Office of Economic Opportunity, founded by President Lyndon Johnson to wage the "war on poverty"; according to its brochure, the Center has been started "to provide technical assistance and grants to low-income projects." It is difficult to accept the claim that "they [the Center] emphasize that what they are looking for is not the development of a 'poor people's technology.' Rather, they would like to see low-income people become leaders in the adoption of technologies upon which everyone must increasingly rely in the future."[1] The image of the American poor leading the American rich contains more pathos than humor. It also points to an important aspect of how many Western societies seem prepared to accept AT—it must be for "them," not for "us." Appropriate technology is proposed for the urban ghettos, for the rural poor, for Navajo Indians, or for Newfoundland fishermen. It is *precisely* as a "poor people's technology" that it is viewed; tragically, the eager proponents are going along with this tendency. Tragically, because poor people, like poor countries, have the greatest vulnerability to error—I will not belabor the patient reader with the millstone analogy once more.

A third group is commonly associated with the Appropriate Technology movement—the youth culture. The effect of the youth culture has been less tangible than that of the environmentalists or of the neo-Utopians, but it is probably more widespread and perhaps more important. This is because the involvement of the youth culture with technology, particularly in California and the American Southwest, actually predates the others' involvement and

[1] "NCAT: Appropriate Technology with a Mission," *Science*, March 4, 1977.

is very much their precursor. It could be argued that the California dream is not usefully related to the infinitely richer theme of world development that this book aims to examine. Nevertheless, many of the technologies that are referred to as "appropriate" appeared in the youth culture in the late 1960s, some time before *Small Is Beautiful*. Their impact on Western society has been largely through books, and as the books of the youth culture described actual experiences, this gave them an immediate impact— that of "true stories"—and also lent a credibility, which they might not otherwise have had, to the later ideas of people such as Schumacher and Illich.

One of the centers for the American youth culture in the 1960s was northern California, and not surprisingly it was there that the youth culture's involvement with technology began. Part of this involvement was due to the natural, and historical, American penchant for technology and technological improvisation, but it was also due to the influence of one man: R. Buckminster Fuller.

## HOME IS A DOME

Richard Buckminster Fuller, engineer-inventor, has been a hero to a number of American generations. His first book, *Nine Chains to the Moon,* was published in 1938 and caused the usually acerbic Frank Lloyd Wright to write in his review: "Buckminster Fuller, you are the most sensible man in New York." Fuller invented a car, the Dymaxion automobile, which, though short-lived, was featured at the Chicago Century of Progress Exposition of 1933–34. In 1945 he designed and built an industrialized house which was totally produced in an aircraft factory, using technology developed during World War II.

All of these inventions had given him a certain notoriety; nevertheless, the vast majority of his projects rarely passed the prototype stage. The device that finally did find widespread application and which made him a public figure was the "geodesic dome." It was the geodesic dome, also, that later formed the link between the ideas of Fuller and the American youth culture of the 1960s.

Fuller's various inventions evolved from an attempt to find the vectoral basis for physical phenomena: what he called "energetic-synergetic geometry." The most graphic example of this approach was the geodesic dome: a structure of octahedrons which used a spherical system of construction—the "great-circle chords." The result was a structural system of maximum economy and strength using the minimum of materials and capable of very large spans. Fuller built the first geodesic dome with students at the Chicago Institute of Design in 1949.[2] There followed a period of further experimentation, often at schools of architecture and design, until the geodesic dome was adopted by the U. S. Marine Corps, which built more than 300 in various locations. The U. S. Air Force used the dome as a radar protective shelter (radome) in its Distant Early Warning (DEW) line installations in the Canadian Arctic. The geodesic dome became widely recognized in the 1950s as a symbol of American know-how: geodesic domes were used as exhibition pavilions by the United States Government in Milan, Kabul, Bangkok,

[2] There is some question as to whether Fuller is the actual inventor of the geodesic dome. Lloyd Kahn, a proponent of do-it-yourself geodesic domes in the 1970s, has pointed out in *Shelter* (Bolinas, Calif., 1973) that great-circle geometry has been known by Southeast Asian basket weavers for some time and that a thin shell planetarium had been built by the German optical firm of Zeiss at Jena as early as 1922 utilizing a dome derived from the icosahedron, a twenty-faced polyhedron.

Tokyo, and Moscow. Throughout that decade, theaters, auditoriums, and industrial buildings were built using geodesic domes, the largest over 300 feet in diameter.

The geodesic domes at this point fell into three categories of use. They were either very large spaces, such as theaters or industrial spaces; were used where speed of erection was a factor, such as military shelters or exhibition pavilions; or were used where great structural strength was required, such as the exposed radome installations. The use of geodesic domes for housing was rare. The U. S. Air Force did build a dome in Korea to serve as bachelor officers' quarters; some small experimental domes were built by students as housing prototypes. But it was generally felt that geodesic domes were advantageous when they were big and that they were not particularly suitable for housing. One of the definitive books on Fuller, Robert Marks's *The Dymaxion World of Buckminster Fuller* (1960), prepared in collaboration with Fuller himself, did not show a single dome used as a home. The notable exception, predictably perhaps, was to be for Fuller himself. In 1963 he built for himself and his wife a thirty-nine-foot-diameter plywood dome in Carbondale, Illinois, which would serve as his home for a number of years. Little did Fuller suspect (or did he?) that only four years later a "city of domes" would appear on the western plains of the United States.

Trinidad was a small, dying coal town in southern Colorado, reputed to have once been the home of Kit Carson, a fabled hunter, scout, and Indian fighter of the Old West. On the outskirts of Trinidad a group of long-haired, buckskin-clad, modern-day pioneers established in 1967 what was to become a mythic countercultural community —Drop City. Drop City, known by the amused but ap-

parently tolerant locals as "Dump City," was a commune
of about twenty artists, writers, and painters. As one of
the founders laconically put it, "We heard R. Buckminster Fuller lecture in Boulder, Colorado, and decided
to build domes."[3] The domes were unlike any that had
been constructed previously. Instead of being built with
machined and molded high-performance materials, they
were made of old lumber, chicken wire, and stucco. A
number of domes were built with used-car tops: "In cutting cartops the first lick is the hardest—if you don't get it
just right the axe bounces off the solid surface—but once
it's started you just work in the previous cut—sort of like a
can-opener the first lick is also the one that makes the
most noise."[4]

The dozen or so domes on "six acres of goat pasture"
at Drop City had a remarkable influence. In 1967 there
was a migration of young people hitchhiking across
America to Haight-Ashbury in San Francisco. Like For-
mentera Island in the Balearics, Lake Atitlán in Guatemala, or Katmandu in Nepal, Haight-Ashbury was a
magnet to the young, the place where one should be—a
counterculture Biarritz of the sixties. Drop City became
quite literally a drop-off point for these transcontinental
travelers, and hundreds of people saw the New Jerusalem
in person. Thousands others read about it in Steve Baer's
*Dome Cookbook* (Corrales, N.M., 1968) and in later
publications. The dome quickly became part of the myth—
"To live in a dome," one of the "Droppers" wrote, "is
psychologically to be in closer harmony with natural
structure." The dome certainly *looked* different, the space

[3] Bill Voyd, "Funk Architecture," in Paul Oliver, ed., *Shelter
and Society* (London, 1969).
[4] Peter Rabbit, *Drop City* (New York, 1971).

inside *was* different, and, if contemporary accounts are to be believed, living in a nonrectilinear environment had a distinct effect on the persons involved. Domes sprang up throughout the West, built not only by individuals, but by other communes, most notably two Drop City offshoots, Libre and Lama, in the foothills of Colorado and New Mexico. And in California a "free school," Pacific High School, began building domes as housing for the students.

Drop City seems to have been the first dome realization that significantly influenced the so-called flower children. However, the first mention of geodesic domes in connection with the youth culture had occurred two years before. According to Tom Wolfe's *The Electric Kool-Aid Acid Test* (New York, 1968), it was Ken Kesey who first wanted to build a geodesic dome in 1965: "For months Kesey had been trying to work out . . . the fantasy . . . of the Dome. This was going to be a great geodesic dome on top of a cylindrical shaft. It would look like a great mushroom."

Ken Kesey was, and is, a writer of some reputation (*One Flew over the Cuckoo's Nest, Sometimes a Great Notion*). For three years, beginning in 1964, he and a group of followers, calling themselves the Merry Pranksters, were the spearhead, the rolling panzer brigade, of the Californization of the American youth culture. Their influence on the counterculture was not dissimilar to the influence that certain groups in the art world (the Bauhaus, de Stijl, the Group of Seven) had, decades earlier, on painting. They innovated, experimented, and led the way. To the extent that northern California was the Paris of the youth culture, the Merry Pranksters were its Impressionists. The Pranksters, like the Impressionists, were exploring the tactile impression of events. "They know *where* it is," Wolfe quotes Kesey as saying, "but they don't know *what* it is."

Though it was never built, it is likely that the concept of Kesey's dome percolated into the psyche of the youth culture, together with such Prankster discoveries as Day-Glo paint, the bus as nomadic shelter, clothing as costumes, and, not the least, psychedelics.

Wolfe described how one of the Pranksters "took some LSD, right after an Explorer satellite went up to photograph the earth, and as the old synapses began rapping around inside his skull at 5,000 thoughts per second, he was struck with one of those questions that inflame men's brains: *Why Haven't We Seen a Photograph of the Whole Earth Yet?*" This Prankster, whose name was Stewart Brand, was so taken with this pregnant question that he addressed letters to all the notable persons he could think of: world leaders, politicians, writers, and thinkers. He received only one reply—from Buckminster Fuller, who answered, quite reasonably, that from outer space you could only see one half of the earth at a time. Later, Brand met Fuller who did concede that *if* it were possible to see the entire globe at once it would indeed alter man's consciousness. "You could say," Brand was later to recount, "that was the beginning of the project." The "project," which would make Brand the Denis Diderot of the youth culture, was a contemporary *Encyclopédie* he called *The Whole Earth Catalog*.

## THE WHOLE EARTH CATALOG

Like Diderot's eighteenth-century enterprise, *The Whole Earth Catalog,* in a series of editions which began in 1968, sought not only to give information but to guide opinion. It dealt primarily—as its later critics were quick to point out—with consumption (this was, after all, twentieth-century America), but rapidly became a vehicle for introducing a range of intellectual ideas to the unlearned young.[5] The youth culture until then had been long on youth but rather short on culture; in many ways, Brand's catalog supplied the latter.

*The Whole Earth Catalog* was inspired, in part, by Leon L. Bean, the inventor of the Maine Hunting Shoe, who established the L. L. Bean company, in Freeport, Maine, a mail-order business catering to hunters and lovers of sportswear and the outdoors. The products that he sells, his own and others, include tools, clothes, and camping equipment which are distinctive because they are chosen for their high quality and on the basis of actual experience. They are generally not subject to changes in fashion (the Maine Hunting Shoe has remained substantially the same since 1912), and they reflect the needs of the user, rather than those of the marketplace. As he prepared for the first edition of *The Whole Earth Catalog*, Brand made a tour of the American Southwest. He returned with the conviction that "people in the communes

---

[5] In *The Making of a Counter Culture*, Theodore Roszak expresses a popularly held view when he writes: "The young, miserably educated as they are, bring with them almost nothing but healthy instincts. The project of building a sophisticated framework of thought atop these instincts is rather like trying to graft an oak tree upon a wildflower." It was romantic to view the youth culture as comprising *enfants sauvages*—and the youth did nothing to dispel this notion—but it was false nevertheless. The youth culture was probably no less educated than American society in general, and probably slightly more so. Indeed, the youth culture's innovators were very *well* educated: both Kesey and Brand had been at Stanford, the former on a graduate fellowship, the latter majoring in biology. Virtually all the "outlaw designers" either had university degrees or had at least spent some time at a university. Many had also spent time in the military, which, whatever it does, is hardly likely to turn out wild flowers.

didn't know what the hell they were doing." The *Catalog,* partially a consumer guide, described a range of tools, outdoor equipment, and do-it-yourself manuals, but it went far beyond that. A list of the chapter headings included "Whole Systems," "Shelter and Land," "Community," "Communications," "Industry and Craft," "Nomadics," and "Learning." Above all the *Catalog* dealt with ideas; more than three quarters of the items listed were books.

Though the New York *Times Book Review* described *The Whole Earth Catalog* as "hip Horatio Alger trading blue chips in the greening of America," it was an eclectic publication that exhibited none of the technophobia of Charles Reich's bestselling book *The Greening of America.* The *Catalog* included reviews of authors as diverse as the futurologist Herman Kahn, Barry Commoner, and Norbert Wiener. It reviewed *Fortune* as well as *Rolling Stone.* It had sources for buying army surplus and aircraft as well as weaving looms and folding bicycles. The technological optimism of Buckminster Fuller, who was featured on the first page, was evident throughout. The standard texts on this period, *The Greening of America* and *The Making of a Counter Culture* (neither of which makes any mention of Fuller), emphasized the antitechnological bias of the youth culture. This is, I believe, a mistaken interpretation. Though it was antiestablishment, the youth culture was not, at first, antitechnological; it was immensely attracted to technology—the technology of backpacking and camping, video and film, music and homesteading—and much of this attraction can be attributed to *The Whole Earth Catalog.*

Diderot spent twenty years writing his *Encyclopédie;* Stewart Brand, hardly in the same league, produced the first sixty-four-page *Catalog* in 1968 in an edition of 1,000 and, only three years later, churned out *The Last Whole Earth Catalog* (447 pages), which sold over 1½ million copies. Part of the success of the *Catalog* was due to Brand's discovery that anyone could be his own publisher and, in effect, printer. The IBM Selectric Composer typewriter and the Polaroid MP-3 copy camera enabled the individual publisher, with minimum investment, to do all book layout himself. Offset printing technology altered significantly the traditional economies of scale. Thus the *Catalog* could grow from 1,000 copies to 100,000, using essentially the same techniques. When it was decided to terminate publishing with one last *Catalog,* there had been enough publicity, and sales, that the large New York publishing houses were interested. From this emerged another discovery: the New York publishers knew how to sell books, but they did not have the editorial (or technical) know-how to produce the kind of book that the youth market was obviously interested in. As a result, *The Last Whole Earth Catalog* was published and printed by Brand in 1971 and then shipped to New York for final distribution by Random House.

Just as the *Encyclopédie* was not produced by Diderot alone, but with the collaboration of people such as Voltaire, Jean-Jacques Rousseau, and Charles de Montesquieu, so the *Catalog* was the result of a group effort and very much a spawning ground for other ventures. Lloyd Kahn, who had co-edited a number of editions of the *Catalog,* published a manual on geodesic domes (actually prepared using Brand's equipment) in 1970 called *Domebook 1,* and a year later he produced *Domebook 2,* which, following the pattern set by the *Catalog,* was prepared in California and distributed by a large East Coast publisher. *Domebook 2* sold a large number of copies, widely promoting both dome construction and the ideas of Buckminster Fuller.

The technological optimism of *Domebook 2* and *The Whole Earth Catalog* had been fueled to a large extent by a meeting that took place in March of 1969. This event took place in an abandoned factory in the New Mexico desert, not far from Alamogordo, where the first atomic bomb had been tested twenty-four years previous. It brought together 150 self-styled "outlaw designers" from Drop City, the Libre commune, Pacific High School, and, of course, the fledgling *Whole Earth Catalog* which reported the event in its March supplement. (Stewart Brand: "If I had to point at one thing that contains what the *Catalog* is about, I'd have to say it was [the Alamogordo meeting].") Most of the individuals concerned had built domes, and in fact they met inside a large, portable geodesic, but the optimistic "agenda," in addition to seminars entitled "Materials and Structure," also covered "Energy," "Man," "Magic," "Evolution," and "Consciousness."

The New Mexico gathering was significant for two reasons. First, it was an identifiable landmark which located a large group of individuals in one place at one time (another landmark event of that year was the Woodstock festival) and had an important influence (by their own admission) on the later careers of the people involved. Secondly, no attempt was made to define or isolate what was happening as a "movement"—if anything, a conscious attempt was made to avoid proseletyzing. Neither was there any evidence at this meeting of the technohysterics that were to characterize later meetings; here no line was drawn between "good" and "bad" or "small" and "big" technologies. The line, if there was one, was probably between talking and doing. "You have to make the change yourself," Steve Baer, one of the organizers, wrote "or you have to shut up about it."

A few years later it was becoming apparent that, indeed, there were coming to be many more talkers than doers, in large part as a result of commercialization of the counterculture.

## THE ALTERNATIVE CONSUMER

In October 1973 two events accelerated the trend toward commercialization: the Organization of Petroleum Exporting Countries (OPEC) more than doubled the price of oil, and the Arab oil producers instituted a total embargo on oil exports to the United States; the so-called energy crisis had begun. This momentous event had a number of side effects which directly influenced the American youth culture and contributed to the kind of atmosphere that prompted public interest (happily catered to by the media) in the subject of energy and, as oil prices kept rising, in ways of getting "cheap" energy. This was particularly true at the individual level, as householders lining up at service stations and paying their heating bills, realized that the energy crisis was affecting them directly and (so it seemed at the time) irrevocably. Quite literally overnight, the previously ignored activities of the American youth culture with windmills and solar energy became "news." An enormous market developed in do-it-yourself manuals. There was a rush to present all sorts of technologies as simple, easy to build, and cheap, whether or not this was in fact the case.[6] The success of *The Whole*

[6] *The Handbook of Homemade Power,* published by the *Mother Earth News* immediately after the energy crisis in 1974, proclaims on its cover: "Heat your home! Use the wind to make electricity! Power a shop, house or farm with a water wheel! Run natural gas

*Earth Catalog* had brought a host of imitators, both good and bad; the energy crisis, whose effect on the psyche of American youth was as great as its effect on the pocket-books of their parents, precipitated the expansion of the youth consumer market from clothes and music into life-style.

The marketing of a life-style, like the marketing of a soft drink, requires gross oversimplification combined with positive user reinforcement. Problems are minimized or left unmentioned; the *image* takes precedence over the *reality; acquiring* is more important than *learning.* "American society has a remarkable capacity to absorb change," Nora Ephron once wrote, "and then turn it on its head." The public acceptance of what was now starting to be called "alternative technology," and which—once the energy scare was over—would mutate with "interme-diate technology" to become "appropriate technology," had turned many of the lessons of the outlaw designers on *their* heads. The protests of Brand and others went unheeded.[7]

appliances—even your car—on methane that you produce yourself! Cook on a wood-burning stove! Yes, there ARE answers to the en-ergy crisis . . . and this book tells you how to make them work for you."

[7] A preview of Brand's *Whole Earth Epilog* (Menlo Park, Calif., 1974), a follow-up to *The Whole Earth Catalog* that met with less commercial success, appeared in *Harper's,* April, 1974. In the "Soft Technology" section Brand wrote: "Welcome to the panacea de-partment: there is no such thing as a panacea. Solar collectors, windmills, wind-generators, biosphere houses, waterwheels, meth-ane: relax and accept some trade-offs—higher independence: higher cost: more work: lower power: lower convenience: higher gut-satisfaction: better balance with environment." Curiously, this statement did not appear in the final version of the *Epilog.* Steve Baer told the *Mother Earth News* in a July 1973 interview: "It's easy for these things to become cults, you know. Like domes have become a cult in the counter-culture . . . and wind generators . . . and so many things. And once they do, people become blinded to design."

The most harmful effect of this commercialization process was the claim of many books and magazines that the youth culture now had "its own technology," a counterpart to its own clothing and its own music. The claim was untrue not only because all three were almost completely the product of large nonyouth enterprises, but also because it maintained that there were different "kinds" of technology, whereas science, technology, and common sense demonstrated the opposite. Windmill rotors follow the same aerodynamic principles as airplane propellers; in some ways they are neither simpler nor less complicated. But the purveyors of people's technology, radical technology, or homemade technology were selling books (above all) on precisely the premise that there *was* a "new technology" (cheaper, simpler, etc.). Public gullibility being what it is, and the fact that other experts were at the same time advocating the "new mathematics," the *"nouvelle vague,"* even the "new left," made it inevitable that this schizophrenic view of technology should take hold.[8] This set the stage for the rather rapid public acceptance of the categorization of technologies into small/big, appropriate/inappropriate, or just plain good/bad.

This fragmentation of technology is quite in opposition to the pronouncements of Buckminster Fuller, who once said, "It is *all* technology." Fuller had felt that the young generation's interest in geodesics could serve as a springboard to further development. His faith in them was formidable, and consequently he devoted much of his time during the late 1960s and early 1970s to addressing the

[8] The single largest entry in the cumulative index of *The Whole Earth Catalog* and *The Whole Earth Epilog* was for items beginning with "new": there are thirty-six entries. For those interested in such illuminating trivia, the runners-up were "black" and "China."

young directly. He spoke on "Hippie Hill" in San Francisco, at colleges, high schools, and less formal gatherings. In 1973, for instance, he addressed groups of 1,500 to 5,000 young persons on 124 occasions (every three days!). "Start doing your own thinking," he told them. "Pay no attention to anybody; start doing what needs to be done." It is not surprising that so many of the individuals who are prominent in this chapter acknowledged a specific, personal contact with Fuller as a major influence.

It was Fuller's hope that this generation would share his "whole earth" view and would be prepared to carry on with the "design science" that he advocated. In an interview appearing in *Domebook 2* Fuller was asked if he thought that there was any conflict between making geodesic domes by hand and mass producing them with high technology. Fuller answered that he himself had experienced the excitement of personal experimentation, "but after you've done it for a while and so you really feel it and understand it, you'll feel that . . . there are more important things to do." He also warned that personal discoveries should not insulate the individual from society at large: "I just want kids not to be disdainful of the other man—in finding yourself you really ought to be finding the other." The youth culture did not seem ready to make this conceptual leap. For one thing, Fuller's optimism was becoming *démodé*. The energy crisis, the Vietnam War, the predictions of books such as Barry Commoner's *The Closing Circle* (New York, 1971), and the Club of Rome's *The Limits to Growth* (New York, 1972), all conspired to create an atmosphere of gloom. Fuller's faith in technology (which he never separated from man) came to be seen as something hopeless or, worse, sinister.

I hasten to point out that the youth culture's involvement with technology had never been as widespread as its

interest in, say, music. Often its use of machines, such as
electric guitars or motorcycles, was more or less uncon-
scious. Nevertheless, unlike the German *Wandervögel,* a
large part of whose time was spent camping and hiking
(wandering) in the countryside, this youth culture was
inextricably a part of technological America. The success
of *The Whole Earth Catalog* had been due to Brand's in-
sight (via Fuller) that modern technology could be used
by the individual for his own ends. The turning away from
Fuller that took place in the mid-seventies, was characterized by a change in this attitude to technology in particular and to modernization in general.

I have already mentioned the gloomy atmosphere of
that time, topped, if that is the right word, by the debacle
of the Watergate scandal that destroyed a United States
President. The tendency in American life that followed
was a search for facile solutions (or facile politicians), a
search that likewise affected the youth culture. The pragmatism of the early years gave way to a desire for straight-
forward beliefs—the good/bad technology syndrome. The
emphasis on personal experimentation was replaced by a
need to order, to define, to organize. In many ways, the
Appropriate Technology movement fulfilled these desires:
it was seen as a complete belief system unhampered by
judgment; there were villains and heroes, there was something to belong to. Disenchanted engineers began, quite
straightfacedly, to call themselves "appropriate technologists." Newsletters, journals, and organizations prolif-
erated. Having created an artificial and unnecessary
fence, it was only important to know, and for others to
know, which side of it you were on. Technique had de-
cayed into belief.

There was more than a touch of nostalgia attached to
this renunciation of the machine. There was a disillusion

with present-day technology and, since the future was obviously conditioned by the present, the tendency was to turn to the past. The admiration for communist China was due in part to the fact that they used old-fashioned techniques (the present *dis*enchantment with recent post-Mao Chinese modernization can be seen as a corollary of this view). Above all, the nostalgia was romantic, for it did not take into account the enormous extent of the modernization process it was attempting to replace.

There were a number of symptoms of the shift. Kahn's successor to *Domebook 2*, called *Shelter*, which he edited in 1973, was an attempt to deglamorize the geodesic dome and to redirect the attention of young builders to more traditional construction methods. *Shelter* was severely critical of earlier dome-building efforts, notably in an article by Kahn called "Smart but Not Wise." Fuller and modern technology likewise came under attack, while picturesque photographs extolled the dubious virtues of Nebraska sod houses and Scottish crofters' cottages.[9]

A more recent publication, *Space Colonies* (edited by Stewart Brand, Sausalito, Calif., 1977) invited a number of "notable people" to comment on a proposal to establish extraterrestrial settlements on artificial satellites. Almost all the notables were either enraged, dismayed, or skeptical. Kesey said it had no appeal for him at all. Schumacher offered his support, on the condition that he could nominate the passengers. Social commentator Lewis Mumford called it "an infantile fantasy." The fact that Russell Schweickart, an Apollo 9 astronaut, and Carl Sagan, a Cornell University astronomer, reacted positively was perhaps predictable, as was probably also the hearty

[9] The romanticism of *Shelter* was replaced by a more realistic pragmatism in its sequel, also edited by Kahn, *Shelter II* (Bolinas, Calif., 1978).

endorsement by Buckminster Fuller. What was significant was Fuller's estrangement from current fashion in taking this position.

Brand's *Whole Earth Epilog,* which followed his *Last Whole Earth Catalog,* contained a new section on "Soft Technologies," adopting a term that had become fashionable in Britain.[10] The British have always impressed their American cousins (unduly, in my opinion) with their glib tongues, and it is no coincidence that the lexikon of AT is British, not American. "Soft," "intermediate," "alternative," "radical," "low-impact," "eco-," and "appropriate" all appeared for the first time in British publications. This can be misleading, for the contribution of the British youth culture was largely cosmetic; the actual accomplishments were mainly American. The focus on solar and wind energy originated in California and New Mexico communes. The first experiments in new types of house construction were likewise American, initially domes but later "handmade houses," the latter closer to AT ideals. *The Whole Earth Catalog* became the model for many publications. There was something faintly ludicrous about a high-powered United Nations bureaucrat holding aloft a publication that had been produced in somebody's garage

10 The use of "soft" has many connotations. Some writers have attempted to relate it to the feminist movement. Others use it in juxtaposition to "hard"—after all, who wants hard if you can get soft? In another but related context, Peter Berger has used the term "soft socialization process" to describe the kind of socialization that takes place within the youth culture and that produces "individuals used to having their opinions respected by all significant persons around them, and generally unaccustomed to harshness, suffering or, for that matter, any kind of intense frustration" (*The Homeless Mind,* New York, 1973). Presumably, if Berger is right, such "soft" persons would be attracted to a "soft" technology. Of course, whether or not such a technology could actually exist is another matter.

and telling an international conference, "Look what the young people are doing. We can learn from this."

Though the AT movement appropriated much of the California dream, the converse was not necessarily true. Steve Baer has been particularly critical. "What *is* Appropriate Technology?" he asked. "I first heard the word from Ron Alward at the Brace Research Institute about four or five years ago and I thought it was great, but when I started thinking about it, I realized I don't know what the hell it is. Every technology is appropriate for something. Just as a designer builds a house out of boards and steel, so the bureaucrat builds out of slogans and jargon, and that is what all the 'appropriate' technology talk is about."[11]

## BUCKY MEETS FRITZ

I would like to close this chapter with an autobiographical recollection. During the 1960s, as an architecture student in Canada, I had been exposed to the ideas of Buckminster Fuller, through both his books and lectures; I had watched his big dome being built for the United States exhibit at Expo 67, the Montreal world's fair, where I was employed as an assistant to the site architect on Habitat, an adjoining building. Later, working as a researcher, I had the opportunity to meet and speak with Fuller; as on so many of my generation, Fuller's influence was both personal and lasting. At the same time, I was aware of, and sympathetic to, the idea of intermediate technology. George McRobie, cofounder of the Intermediate Technology Development Group had visited McGill University in 1972 and we had appeared

[11] Interview with Sandra Oddo, *Solar Age*, January 1978.

together in a short film on technology and development.

Cultural heroes tend to live in different, and insulated, worlds. They rarely confront each other; after all, they are supposed to have the final word and confrontation might lead to disagreement. Nevertheless, there is a morbid interest in such meetings, as evidenced by the popularity of an event such as the Kennedy-Nixon presidential debates in 1960. What *would* Norman Rockwell have said to Norman Lear? Or, to return to the subject of this chapter, what sort of palaver would Buckminster Fuller have had with E. F. Schumacher?

I witnessed a meeting of these last two on February 26, 1974, at the United Nations.[12] It was not, of course, a "gunfight at the UN corral." The rarefied atmosphere of the United Nations discourages arguments, and in any case Schumacher and Fuller spoke consecutively, to an audience rather than to each other. But it was a confrontation of sorts; the only time, to my knowledge, that the two met.

The contrast between the two figures was striking. Schumacher, the British civil servant, the survivor, spoke about reducing the scale of technology, addressing the problems of the poor, his dry, self-depreciating humor thinly masking an apparent disillusionment with progress and modern life in general. "I belong," he said by way of introduction, "to what is undoubtedly the lowest form of life on earth. I am an economist." His starting point was a perceived duality in the world: "One finds, of course, that the scientists and technologists of the rich countries work

[12] Schumacher's and Fuller's statements are from tapes of a meeting organized by the Preparatory Planning Group for the Vancouver Habitat Conference on Human Settlements, part of the United Nations Environment Program. This meeting was related to the UN Conference on Human Settlements held in Vancouver in 1976.

on the problems of the rich countries, and the scientists and technologists, far less numerous, of the poor countries work on the problems of the rich countries also." There was also, according to him, a duality in technology: ". . . it struck me very forcefully that there exists a low-level technology which is too low to provide a decent living; and that international aid, and of course also private enterprise, was infiltrating into these countries a very high technology which may fit in the main cities, but does not fit the big rural areas where about 80 or 90 per cent of the population live; and there is nothing in between." Schumacher spoke of rich man's technology and poor man's technology. All in all, he omitted many of the contentious issues raised in his book *Small Is Beautiful* (which had been published the year before) and limited himself to describing intermediate technology.

Buckminster Fuller spoke, as he often does, for an extended period of time. Though he was at this time seventy-eight years old, it was the first time he had been invited to the United Nations, and one could sense that he did not want to waste the opportunity. Using maps, he went over the twentieth-century developments that had resulted in a shrinking world; he described it as a "one small-town world." In such a world one must "think about things in the biggest way possible." "Of course we're in trouble," Fuller acknowledged, because the systems of accounting, of building, of making decisions were all inadequate to the new conditions. A threshold had been reached, and "repairing yesterday" would no longer suffice.

As I listened, I realized that there were warnings contained in much of what Fuller was saying. He was concerned that technology was being misunderstood. On recycling: "You must always think about what is the task to be done. You certainly don't design an airplane or a radio set by asking yourself how am I going to use up waste material that nobody wants." On self-reliance: "*All* of humanity is dependent on *all* resources." On labor-intensive technology: "I found a really great worry in looking at your agenda where I saw you talking about . . . using the waste labor . . . in the first place I don't want to ever think about labor that way any more. *Man is not here for the muscle work at all; he is here for his head work* [my emphasis]. And so I'd like to just get completely rid of the idea of how to make employment, which has been one of the great political games up to now, where people are then beholden to the politician for the job that he arranges they get."

"*Man is not here for the muscle work at all; he is here for his head work.*" In a single sentence Fuller had touched on a vital distinction. The late political theorist Hannah Arendt characterized two aspects of the human condition which she referred to as *animal laborans* and *homo faber.* The former represents those aspects that relate to the physical world, the biological processes of the human body, and especially to the natural world that surrounds man. *Homo faber,* in this characterization, represents those aspects that deal with the artificial world that is created *by* man. Arendt gives a description of *homo faber,* man the maker, in *The Human Condition* (Chicago, 1958): "his instrumentalization of the world, his confidence in tools and in the productivity of the maker of artificial objects; his trust in the all-comprehensive range of the means–end category, his conviction that every issue can be solved and every human motivation reduced to the principle of utility . . . his equation of intelligence with ingenuity . . . finally, his matter-of-fact identification of fabrication with action."

It is quite obvious that Buckminster Fuller incorporates
many of the characteristics that Arendt ascribes to *homo
faber*. It is perhaps less obvious that E. F. Schumacher
reflects many of the concerns of the other side of the
human condition, that of *animal laborans*. Whereas for
Fuller the world could not "run out" of materials since
these were the result of human fabrication, not of nature,
for Schumacher, nature was the "provider of all good
things" and what it had "given" it could "take away."
Unlike Fuller, Schumacher did not identify with modern
technology; if anything, he felt threatened by it. Was it a
coincidence that whereas Fuller's recreation is sailing (his
boat, a sophisticated racing sloop is called *Intuition*),
Schumacher listed "gardening" as his hobby in the British
*Who's Who?*

Perhaps I am making too much of the *animal
laborans/homo faber* and Schumacher/Fuller analogy.
Arendt's distinction was, after all, a philosophical one and
not to be taken literally, and the meeting of these cultural
heroes was, finally, accidental and of slight historical importance. But the parallel seemed to me then, as it does
now, too strong to ignore. The struggle between low and
high technology, the evident contrast between Schumacher and Fuller, the dualism that permeates AT—all
are rooted in a deeper conflict. As far as Schumacher and
the Appropriate Technology movement was concerned,
man was not here for the head work, he *was* here for the
muscle work, or at least it was the muscle work that permitted him to be here. Fuller wanted man to do what
needed to be done; Schumacher wanted him to be happy.

As I listened to Schumacher, and then Fuller, speak, I
realized the California dream was over. I have gone on for
some length about this encounter because it was a kind of
*satori*—what is, in Zen Buddhism, "a moment of enlight-

enment," or what Jack Kerouac called "a kick in the eye." It strengthened in me the suspicion that some aspects of Appropriate Technology were not an extension of earlier work but, perhaps unconsciously, a repudiation. Someone once asked me, "What side of the fence are you on?" It had not occurred to me then that there was a fence. I realized in that United Nations meeting room that the fence did indeed exist (though not where I imagined it might) and that I was beginning to know on which side of it I stood.

*Chapter 5*

# TECHNOLOGIES IN CONTEXT

GLENDOWER: "I can call spirits from the vasty deep."
HOTSPUR: "Why, so can I, or so can any man;
  But will they come when you call for them?"
                              —WILLIAM SHAKESPEARE,
                              *Henry IV, Part I*

I have traced the development of the AT movement from
its origins in the youth culture as well as in the work of
E. F. Schumacher. I have also examined its relationship to
the less developed countries, both in a positive and in a
negative way. The focus of the previous chapters has been
on technological theory, but, though it is often claimed
that "appropriate technology is an approach, not a
specific package of technology,"[1] an examination of AT
must, sooner or later, deal with the machines and tools
that are considered to be "appropriate."

It is quite reasonable that this should be so. Appropriate Technology refers not just to good engineering but
to a specific technology that reflects a new set of technical
and social criteria: small-scale, cheap, easy to use and understand, and encouraging self-reliance and nonviolence.
If present-day technologies do not exhibit these characteristics and are inappropriate, the implication is that a

[1] Ken Darrow and Rick Pam, *Appropriate Technology Sourcebook* (Stanford, Calif., 1976).

new technology could follow these criteria. I will leave the discussion of whether it is possible to develop a completely new technology until the next chapter; for the moment it is sufficient to recognize that this argument posits the existence of such a new technology and that a major part of its success has been its ability to produce examples of such new devices.

What are these machines? It is not difficult to produce a list: various solar devices, almost all hand tools, bio-gas digesters, wind machines, greenhouses, various pedal-powered machines (including, of course, bicycles), composting toilets, and so on. The origin of these devices is largely either from less developed countries (what used to be called "village technology") or from the youth culture. The categories are not hard and fast; hand tools are preferred over machines, but small machines are preferred over big machines, and even big machines are viewed more favorably than very large plants. Thus, in examining any specific technology too closely, one runs the risk of finding out either that the connection to AT is marginal or of falling into a tautological trap—that is, of considering that only successful appropriate technologies are really "appropriate."

If AT is to be taken seriously, if it is to take itself seriously, it must be able to modify itself on the basis of experience. If Appropriate Technology is unable, or unwilling, to do this, it will founder—like so many protest movements—in a sea of self-delusion. This will lead to the situation where, to paraphrase Jean-François Revel, an appropriate technology is not required in the short run to do better than other technologies; it is in any event superior because it is a *good* technology, while the others are *bad* technologies. If criteria of efficiency, productivity, economics, profitability, and (above all) common sense are abandoned, there is a risk of reducing all to simply the ideological level. Perhaps this is happening already. The crucial question should be "Does AT work or doesn't it?"

Carl Riskin has written in the context of intermediate technology: "[It is] a general argument buttressed with specific examples, it cannot be accepted or rejected *per se*, but must be evaluated in the context of the pre-existing situation as well as the desired speed and scale of change."[2] Riskin points out the crucial factors that should be considered when describing a technology: the economic and social circumstances that surround the technology—that is, the historical factors—the rate of change or development, and the scale or extent of development. All these constitute the context for technological development. Thus, in order to learn something about a particular technology, it is necessary to look at that technology in a specific context, just as to learn something about a political belief, it is necessary to examine its application in a particular country.

There are a number of difficulties in identifying small-scale technologies in action. In spite of the fact that a recent World Bank computer search turned up over 9,000 titles dealing with "appropriate technology," there are not many studies that deal in detail with the social and economic successes (or failures) of appropriate technologies. The reasons for this dearth of data are multiple. First, the concept is new and hence most projects are too small, and too recent, to warrant conclusive study. At the same time, the resources of most AT groups are limited, and hence follow-up studies are rare—most documentation describes only the installation process, not later performance. Finally, as I have shown, much of this activity takes place

[2] "Small Industry and the Chinese Model of Development," *China Quarterly*, No. 46, April/June 1971.

under the guise of bilateral aid, where diplomacy tends to replace science. Scornful references by Third World bureaucrats to "rusting windmills" hint at shortcomings, but documentation of such failures is virtually nonexistent.

I have not included any proposed or speculatory projects. These make intriguing and exciting reading—some have been reproduced so often that they might be considered "real"—but paper Utopias cannot substitute for actual experiences. For somewhat the same reason, I have not dealt with what are called "demonstration projects." The aim of demonstration projects is to show, on a small scale, what might be possible on a large scale. They are usually sponsored by organizations (the "demonstration project" is a United Nations invention) and are frequently designed as heavily financed, public relations gestures, often large pedestals for small statues; as a result, such initiatives tend to be cut off from their surroundings and frequently have limited scientific value.[3] The public success of Schumacher's *Small Is Beautiful* and the visual appeal of devices such as wind machines and solar hardware has created a tendency among governmental and international organizations to promote AT demonstrations in lieu of supporting serious efforts.[4] This is not necessarily the

[3] Dr. Michael McGarry, of the Canadian International Development Research Centre, describes a visit to a UNICEF demonstration center of appropriate technologies for farmers outside Nairobi: "A myriad of gadgets are on display; only a few are really relevant. Unfortunately, management of the Unit is UNICEF and not Kenyan dominated. Less than 5% of visitors to the Unit have been Kenyan farmers; the vast majority of visitors have been international travellers representing the UN, other development agencies, and local bureaucrats from Nairobi." McGarry, "Appropriate Technology in Civil Engineering," paper presented at the 1977 Annual Convention of the American Society of Civil Engineers, San Francisco.

[4] Joseph Hanlon, a British science writer, visiting the Centre for

# 115

fault of the practitioners, who are often pawns of a political game.[5]

The Spring 1975 issue of *CoEvolution Quarterly* contains a revealing letter by a bio-gas enthusiast who had embarked on a transcontinental journey to visit a number of prominent author-inventors in the field of bio-gas technology. He reported that of all the digesters he saw, only *one* actually produced gas and that he had the impression that the majority, including some that had been the subject of do-it-yourself pamphlets had never worked at all. Nevertheless, and in spite of the fact that this technology is unsuited to temperate climates (except under certain conditions), bio-gas continues to be described as a universal solution.

A recent project in Prince Edward Island, sponsored by the Canadian government as a demonstration for the 1976 Vancouver Habitat Conference on Human Settlements, has received wide publicity as a "real working solution." In some ways it is that, but it is also a highly subsidized example of "autonomy" which is neither a total success technologically nor an effective demonstration locally.[6]

Alternative Technology in Machynlleth, Wales, describes windmills, a bio-gas plant, and an energy conserving house, but adds that "many of the devices work poorly or not at all." A member of the AT group, which is supported by industry, admits that the windmills are "more touristy than practical." See *New Scientist,* October 27, 1977.

[5] A flagrant example of this was the installation of a much-publicized solar heater at the presidential grandstand in Washington, D.C., during Jimmy Carter's inauguration in January 1976. Less publicized was the fact that the heater refused to function and that presidential lap robes saved the day.

[6] This expensive building, referred to by its designers as "the Ark," has had little impact on the local population, which has minimum access, physically, intellectually, or economically to the

Another project that is supposed to be a successful example of AT was the promotion of local self-reliance through urban gardens in a poor district of Washington, D.C. There are indications that although this project has attracted wide attention in the press, it has been largely ignored by the local population, which was ostensibly to comprise its clientele.[7]

I have tried to choose examples of technologies which are either commonly associated with AT or that exhibit AT criteria and which have been studied in a particular context and on a broad scale physically, but especially over time. My purpose is not to describe the technologies themselves—this has been done many times before—but to throw some light on the economic and social aspects of these technologies. Are the social and economic effects of appropriate technologies predictable? Are they inevitable? And, in a more general way, are all the criteria realizable in practice?

This is *not* an exercise in technological jurisprudence. In spite of the popularity of "technology assessment," I do not believe that it is possible to reach any final decision on the effects of a technology. As Langdon Winner,

project. Rightly or wrongly, the islanders exhibit a high degree of antipathy for what they regard as "a group of Americans heaving money out of the government" (*Macleans,* May 1, 1978). For a variety of reasons, some political and some technical, the provincial government took over control of this project in March 1978.

[7] It is reported that weekly meetings were attended primarily by other experimenters from various parts of the country, not by the local "community" (*Rain,* May 1978). An urban gardening project in Montreal, with which I was involved, was more successful; this was largely because of the fact that the initial goal of self-reliance was abandoned, the technical aspects (greenhouses, hydroponics, etc.) were played down, and the project responded to the actual expressed desire of inner-city people, which was for gardening as leisure, not as cultural secession; see Ron Alward et al., *Rooftop Wastelands* (Montreal, 1976).

an assistant professor of political science at the Massachusetts Institute of Technology, has indicated in *Autonomous Technology* (Cambridge, Mass., 1977), the whole point about unintended consequences is that they are *not* intended, whether or not they are "good" or "bad." Likewise, the effect of actions, human or nonhuman, technological or natural, is impossible to foretell since the effects continue indefinitely, as the famous rhyme about the battle lost as a result of a missing horseshoe nail illustrated.

## BIO-GAS IN INDIA

This Baedeker of AT begins in India. The Indian experiences with AT merit serious study for a number of reasons. First, there is the continuing presence of Gandhi, the nemesis of modernization and an early proponent of a populist approach to the use of technology. Gandhi's influence is not only historical but also contemporary; his ideas have consistently played a role in Indian politics since his death in 1948. This influence was felt in a series of five-year plans from 1950–65, an important component of which was the Community Development movement, a village-level, government-organized program of development. The scale of this movement rivaled the Chinese Great Leap Forward: by 1965 over half a million villages were involved, about 300 million persons. But the movement did not have the desired effect, in spite of the enormous effort involved. It was not successful in eliminating the social differences that existed in the villages, nor was a government-regulated program easily integrated into the traditional, closed communities. Perhaps because it was too ambitious and perhaps because it underestimated the problems involved, the Community Development move-

ment has been generally judged to have been a failure.[8]
The symptom of this failure, as in the Great Leap Forward, was an inability to increase food production, characterized by reduced harvests.

The Community Development era did result in the establishment of a number of organizations which were to
deal specifically with what was then called "village industry" and later came to be called "appropriate technology." These groups have continued since their establishment in the 1950s and have been active in developing
and popularizing a number of technologies, foremost
among which is bio-gas.

There are two widely held misconceptions about
bio-gas technology: (1) that it represents a new and un-
tried technique (for producing power from agricultural
wastes),[9] and (2) that it is a solution to the problems of
the rural poor. Neither of these statements is true.

The production of a combustible gas containing methane (also referred to as marsh gas, dung gas, gobar gas,
or bio-gas) from a mixture of cow dung and farmyard
manure was pioneered in Germany, where a number of
large plants were producing it during World War II,
particularly when petroleum was in short supply. At the
same time, government institutions in India began developing smaller, domestic-size bio-gas digesters for even-
tual use on farms. Although the German work was more
or less discontinued after the war, by the 1950s a fairly

[8] See, for instance, Charles Bettelheim, *India Independent*, trans.
W. A. Caswell (New York, 1968).
[9] For instance, "Another very important fuel which has not been
adequately developed or exploited is methane . . ." (R. J. Congdon, ed., *Introduction to Appropriate Technology* [Emmaus, Pa.,
1977]); or, "A further ecologically sound source of energy which
. . . has been little used in the past, is methane gas" (David Dickson, *Alternative Technology* [London, 1974]).

## FARMING TECHNOLOGY

There have been many attempts in the less developed countries to improve agriculture. Rice fields are generally harrowed with ox-drawn plows in Indonesia, as in most of Southeast Asia.

18

Hand-operated 5—6-horsepower cultivators have been developed at the International Rice Research Institute in the Philippines, though they have not yet found widespread application (18).

Small tractors (19), hardly very efficient with their caterpillar blades, have been developed in China.

19

Hand threshing of rice is widely practiced in the less developed countries, as it was in preindustrial Europe. Even such small mechanical threshing machines as this American model represent an enormous advance in productivity (20).

20

21

Hand grinding incorporates many of the characteristics of preindustrial agricultural practices: a minimum of technology, low productivity, and a high degree of physically debilitating human labor. The hammer mill represents an "intermediate" level of milling technology (21), but can it compete with the roller mill, a "modern" machine invented in the 1870s?

# WINDMILLS AND WATER HEATERS

Most technologies combine low, high, and medium levels of complexity. The sail windmill is a low-power device that can still be found in places such as Majorca (22). It has been suggested that sail mills would be suitable for developing countries,

22

23

and contemporary versions, such as this one designed by Hans
Meyer (23), have been tested, though not widely used. The
traditional fan mill can only be manufactured in countries with

24

a solid industrial base, such as Australia or South Africa, and
the airscrew-type Brace windmill (24) likewise requires a relatively high level of technological accomplishment. A medium-

**25**

level example, more powerful and durable than the sail mill
and less sophisticated than the fan mill or airscrew is the vertical-axis rotor from Ethiopia, shown here with its inventor,
Armando Filippini (25).

26

The thermosiphoning water heater, such as this from Australia (26), may be quite inaccessible to the urban, not to mention the rural, poor. The pillow-type solar water heater represents a much cheaper solution. The one in the photograph being held by its designer, V. S. Nataraj, is made out of a plastic garbage bag and costs less than $2.00 (27).

27

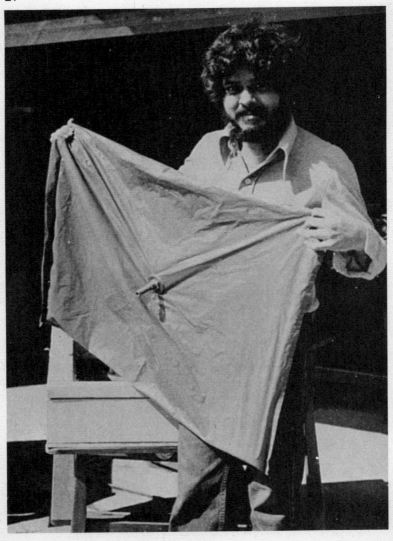

## COMPOSTING TOILETS AND SOLAR HEATING

The apparent success of the North Vietnamese composting toilet was due to a comprehensive approach to the problem of rural health, a social rather than technological strategy (28).

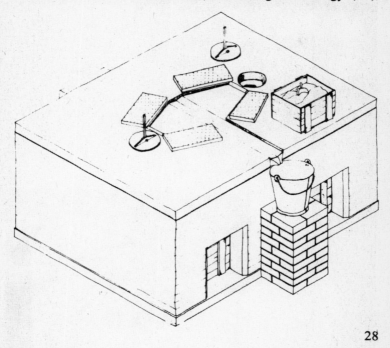

28

Composting toilets can be as simple as the Matson/Warshall drum privy developed in northern California (29), or as complex as the Bioloo, a product of that mythic land of ecotechnology, Sweden (30).

The existence of low-, medium-, and high-technology solutions
to solar house heating points up the danger of oversimplifying the distinction between technologies. Active solar-heated
houses, such as this one in Quebec (31), are rapidly becoming
a status symbol for the relatively wealthy. Passive solar heating
requires considerably less investment, though performance is

considerably less efficient. It is probable that solar-tempered houses, such as this example from eastern Canada (32), which uses a south-oriented greenhouse to trap the heat of the sun, will provide the solution that the vast majority of people can afford.

32

# EARTH BUILDING

Earth building exhibits varying degrees of complexity, sophistication, and cost. It is particularly cheap when used as an infill with bamboo or wood, as this example from Malawi illustrates (33). A more expensive technique requires the addition of

33

cement to earth. This mixture can be used with a manually operated block-making press, the Cinva-Ram, actually developed in Colombia but shown here in a Liberian village (34).

34

large number of bio-gas plants began to be built in India. Several Asian countries followed India's example, and it is estimated that at present there are about 36,000 plants in India, 27,000 in South Korea, and anywhere from 80,000 to 200,000 in China. Thus there is extensive experience in using bio-gas technology for the last twenty-five years, and, in many cases, there is extensive scientific documentation as well. Unfortunately, this documentation does not support the view that "biogas will have an obvious positive impact on the development of the rural areas, which is a vital factor in reducing rural migration to urban centres," as the Canadian *Handbook on Appropriate Technology* (Ottawa, 1976) claims.

Dr. S. K. Subramanian, an Indian engineer, has prepared an exhaustive study of bio-gas technology in Asia, which includes its impact on countries of Southeast Asia as well as on India.[10] He observed a number of characteristics of bio-gas technology: there had to be a minimum number of cattle to provide dung, there had to be land available for disposal of the sludge, there had to be already some significant use of fuel to warrant (financially) the "savings" in switching to bio-gas, and, finally, there had to be capital available for the construction of the plant, which, in 1978, cost between $100 and $200 for a small domestic model. It was Subramanian's conclusion that "on the Indian experience, people who have so far been able to benefit from biogas plants have been in or above the middle-class levels." This was partly due to imbalances in the Indian rural economy, as critics have pointed out, but it was also due to limitations inherent to

[10] S. K. Subramanian, "Biogas Systems in Asia: A Survey," in Andrew Barnett, Leo Pyle, and S. K. Subramanian, *Biogas Technology in the Third World: A Multidisciplinary Review* (Ottawa, 1978).

the technology. The bio-gas digester produced large amounts of sludge, rich in fertilizer value but difficult to transport. Hence only farmers who have large land parcels *on which they live* could readily use the fertilizer. Also, the economic viability of the bio-gas investment was at least partly based on consuming the gas, which presupposed preexistent high levels of fuel consumption, which was unlikely to be the case with very poor families.

Another factor should be considered when looking at the impressive number of installations in India and South Korea (China is a separate case).[11] To a very large extent, bio-gas plant construction has been the result of external inducement. In India, credit of up to 100 per cent was given until fairly recently, and in South Korea between one third and one half of the construction cost was subsidized. Subsidies have now been either reduced or withdrawn in both countries, with a resulting drop in bio-gas installations.

But there are also technical problems. Bio-gas production is proportional to temperature, and falls drastically during the winter. Thus, in South Korea, where plants are simply shut down from December to March, bio-gas can provide only 3–6 per cent of the house heating, though it does provide up to half of the cooking fuel. The Planning Research and Action Institute in India began its bio-gas program with thirty installations. After one year, less than five of the plants were working; the rest had been aban-

[11] There is very little documentation on social and economic implications of the Chinese experience with bio-gas. A useful source for technical data is M. McGarry and J. Stainforth, eds., *Compost, Fertilizer and Biogas Production from Human and Farm Wastes in the People's Republic of China* (Ottawa, 1978). On the basis of India's experience, I would hazard a guess that the majority of bio-gas use in China would be by the *richer* communes and village cooperatives.

doned either because insufficient gas was produced or not enough cow dung was available.[12] Later work attempted to make smaller digesters, but production of gas was still irregular, soaring in the summer and falling in the winter.[13]

The reason for choosing to use bio-gas has varied in different countries, but there was no indication that classical economic theory was being discarded in favor of some new development theory. Though in some countries the value of the gas predominated, in others it was the value of the fertilizer sludge, and even convenience or public health advantages; in all cases, the final judgment was based on the best return for the investment. Likewise, the local economic environment played a big role: in South Korea, bio-gas technology was encouraged by state grants and loans; in India, the high cost of kerosene after the 1973–74 energy crisis encouraged construction of bio-gas plants (over 70 per cent of present plants were built after 1973).

But bio-gas technology has also had some unintended effects. In general, since bio-gas plants have been built by farmers from the middle class and above, economic differences have been exacerbated and not mollified, much like what happened during the "green revolution." More specifically, since cow dung has acquired a higher value, it is no longer freely available to the very poor, who previously used it as cooking fuel.

[12] M. K. Garg, "The Upgrading of Traditional Technologies in India: Whiteware Manufacturing and the Development of Home Living Technologies," from N. Jéquier, ed., *Appropriate Technology: Problems and Promises* (Paris, 1976).

[13] Bio-gas production in India may exceed demand in summer months. If no storage capacity is available, as is usually the case, the extra gas is simply burned off, often at night. I am indebted to my graduate student Mahendra Shah for this personal observation.

In *India: A Wounded Civilization* (New York, 1977)
V. S. Naipaul described attempts to modernize traditional
devices: "After three thousand or more backward years
Indian intermediate technology will now improve the bullock cart." What have been the results? In December
1976 Firestone-India announced a new product: a bullock-cart wheel fabricated from steel with a solid rubber
tire. There are 13 million bullock carts, and their carrying
capacity will now be increased by 50 per cent; of course,
the cost of the new tires is more than 50 per cent higher
than that of the traditional wooden wheel. It is too early
to know definitely, but it is likely that the high cost "puts
the bullock cart beyond the poor peasant who now uses it.
He is worse off if bullock carts are modernized."[14] The
upgrading of indigenous technology often means that
craft manufacture is displaced by industrial or semiindus-
trial manufacture and, as in China during the Great Leap
Forward, when small-industry was established by siphon-
ing off the physical and human resources of the existing
craft industries, the new improved bullock-cart wheel displaced the traditional wood-wheel craftsmen.

The Indian approach to appropriate technology is
heavily larded with polemic, a revived Gandhism encouraged by European visitors. Research groups have
sprung up in a number of locations, though their impact
has not been significant, partly because of a lack of support from the Indian establishment and partly as a result
of their own isolation from the rural areas and their prob-

[14] V. Vyasulu, of the Indian Institute of Management, quoted by
Joseph Hanlon in "Does AT Walk on Plastic Sandals?" (*New Scientist,* May 26, 1977). According to Hanlon, the main motivation
for the rubber bullock-cart wheel was a glut on the rubber market.
More to the point, many Indian bullock carts already have rubber
tires and have had for some time.

lems. An article in the *New Scientist* (June 2, 1977) described a windmill built by one of these university groups:
the sails were made of jute; all the materials, except for
the bearings, were available in any small Indian town; it
operated a water pump that was ingeniously made from a
modified scooter tire. But how useful is such a windmill
project? Except for the coastal areas, there is no wind at
all in India from October to February. The estimated cost
of the windmill and pump are similar to that of a small
bio-gas plant, which experience has shown cannot be
afforded by any but the richest farmers. The device,
though it will probably receive praise and coverage in the
Western press, is not likely to make much impact on rural
India.

AT in India over the past two decades has been seen as
a *cause célèbre*. The teachings of Gandhi and the pronouncements of Schumacher have been taken literally:
"Appropriate technology is a good technology; it will be
successful because it is good." Unfortunately, it has not
turned out that way. Two small examples: small-scale village production of aluminum dishes and utensils received
government support. A few years later a large manufacturer began producing anodized aluminum products, easier
to clean and more scratch-resistant. The village-level aluminum industry collapsed in six months. Soapmaking has
been part of village self-reliance programs since the
1940s. The problem has been that the very poor do not
buy soap, and those who can afford it are prepared to pay
more for the nationally advertised, high-quality product.

Lest the reader get the wrong impression, I hasten to
add that these failures are not necessarily the failures of
all small industry in India. The Indian government has
promoted small-scale industries in various modern sectors

with considerable success. These are as impressive, in their own way, as the more publicized Chinese examples and, like the latter, are approached with a good deal of pragmatism and the emphasis on productivity rather than job creation. Unlike the village technology sector, they are characterized by a minimum of ideology, which may well account for their success.

The evidence so far indicates that AT in India has been a failure for two reasons. First, technical and economic: the AT devices have not performed satisfactorily, have often been too expensive, have produced unmarketable goods, or have been concerned with marginal problems. There is an important lesson here. Good intentions cannot replace good science; it appears that in India polemical considerations and influence from abroad have outweighed common sense.

The second reason for the failure of AT in India is more complex. It resides in the belief that social reform can come about as the result of technological innovation. There is nothing in the Indian experience that supports this view. In *The Challenge of World Poverty* (New York, 1970) Gunnar Myrdal wrote, "Better seed grains can certainly not be a substitute for agrarian reform," and this could be paraphrased as *Better technology (of any kind) can certainly not be a substitute for social reform.* Landlordism, powerful rural elites, conservative banks, and rapacious moneylenders all conspire to maintain the poverty of the landless peasants. These social and political problems require social and political solutions; it is both presumptuous and naïve to believe that technology alone will have any effect in a situation such as this.

It is a mistake to confuse social *change* with social *reform*. Technology is likely to influence the former; it

rarely effects the latter. The telephone, for instance, is undoubtedly an instrument for social change: distances are
reduced, new relationships result, face-to-face contact
may be eliminated in certain activities. On the other hand,
*who* gets a telephone is a function of a social system. The
AT experience in India shows that technology will effect
social changes, but only within the existing social system.
The well-meaning hope that technology could effect social
reform without, or in spite of, the political environment
has proved to be stillborn.

A paper presented at an AT symposium in 1978 asked
the question, "After you've searched your soul for an appropriate technology, how do you get people to use it?"[15]
The AT movement has traditionally concerned itself with
the first part of that question and has thus been accused,
with some justification, of being a "technical-fix" approach. It is quite true that somehow people must be con-
vinced to use the technology, but it is likely that this is
putting the cart before the horse. It is much more probable that successful AT would complement, and emerge as
the result of, a program of social reform.

## COMPOSTING TOILETS IN
## NORTH VIETNAM

An example that illustrates the relationship between intermediate technology and social planning is that of the
composting toilet and rural sanitation in North Vietnam.
This innovation took place in the 1960s, but has only

[15] Allen Jedlicka, "Delivery Systems for Rural Development,"
paper presented at the National Meeting of the American Academy for the Advancement of Science (AAAS), Washington, D.C.,
February 23, 1978.

recently come to light in the West.[16] The North Vietnamese composting toilet originated when peasants who were using human excreta as manure found that composting reduced the smell and improved the fertilizer value of the manure. More importantly, this practice also reduced the spread of infectious diseases. A container was devised with two compartments which served alternately as receptacles for defecation and for composting. This device, which could be built by the peasants themselves from a variety of materials (stone, earth, concrete), played an important role in a widespread program to improve public health in the villages. It is reported that over a five-year period virtually every rural household constructed such a composting toilet, a remarkable achievement, given the fact that rural latrine programs in many less developed countries have consistently failed.

It is tempting to focus on the technological device, but, in *Health in the Third World: Studies from Vietnam* (Nottingham, 1976), Dr. Joan McMichael, a British expert on public health, makes it very clear that the composting toilet installations were only a small part of an integrated program for improving health in the rural areas. The vehicle for social change was the organization of rural paramedical workers, based on the Chinese model, and the technology, which also included improved wells and bathrooms, was developed as the result of this social effort. Social development was not the result of choosing a par-

[16] The North Vietnamese composting toilet was first brought to my attention in 1974 by Dr. Krisno Nimpuno, an Indonesian architect. It is also described in a 1968 booklet published by the Department of Hygiene and Epidemiology of the Democratic Republic of Vietnam. The North Vietnamese composting toilet is similar to the Gopuri composting latrine developed in India during the 1950s.

ticular technology as Schumacher claimed in *Small Is Beautiful,* but rather the other way around.

The North Vietnamese composting toilet was developed after a fairly long period of experimentation which included many setbacks. The fact that the reuse of human waste as fertilizer was already a tradition throughout Vietnam, as it is in a number of other Asian countries, is important to bear in mind. In other cultures, taboos are strong, particularly with regard to defecation, and the acceptance of such a device elsewhere is by no means certain. It is unlikely that the Vietnamese experience will serve as a model for *every* country; there are too many differences, not the least of which is the fact that North Vietnam was a country at war and thus had a highly politicized and disciplined rural population, welded together (like the British in 1940) by enemy aerial bombardment. Nevertheless, it is an example of the relationship that should exist between social action and technology. Technology can only solve technical problems—in this case containing and sterilizing human wastes, thus reducing disease. The apparent success of this device was due to a comprehensive approach to the problems of rural health and a program which was in essence social, not technological.

## WINDMILLS AND WATER HEATERS

One of the basic tenets of AT—and a useful one—has been the idea that for every technological problem there exist solutions of varying complexity and cost, or, as the economists say, solutions that optimize different factor costs. Some solutions minimize labor, some minimize capital investment, some reduce material costs, and so on. This is sometimes misunderstood to mean that there are entire families of devices—windmills, for instance—that represent an intermediate technology. This is an over-simplification, and a misleading one. If one uses the rather simple categories of low, medium, and high to characterize, in a very general way, the cost or level of sophistication of a device, one finds that many technologies combine all three classes.

Windmills, or more accurately wind machines, are a case in point. Although sophisticated and very large wind machines do exist for the purpose of generating electricity, their most widespread use is for driving mechanical devices, especially water pumps. But even water-pumping wind machines need to be regarded selectively. A low-technology example is the sail mill, of which thousands are in use on the Greek island of Crete, as well as in Indonesia. The sail mill uses cloth blades, is relatively feeble and hence used primarily for pumping shallow wells. A high-technology wind machine, often referred to as a fan mill, was a familiar feature of the American rural landscape in the nineteenth and early twentieth centuries. Fan mills are still widely used in South Africa and Australia and, to a lesser extent, in parts of the United States. The fan mill is able to pump wells that are as deep as 1,000 feet, and can be manufactured only in a country with a solid industrial base. It may cost ten times as much, or more, than a sail mill. A medium-technology wind machine falls, obviously, somewhere between the two. One example is Armando Filippini's vertical-axis wind machine being developed in Ethiopia; another is a model from Arusha, Tanzania. These types of machines are characterized by simpler construction, lower efficiency, and lower power than the fan mill and considerably lower price.

It is probable that in many rural poverty areas it is the

medium-technology windmill which will be most useful in providing community water supply. The high-technology fan mill is simply too expensive and, being in most cases an imported machine, usually represents a too complex level of technology. The low-technology solution, such as a sail mill, seems likely to be limited in application; it is still too expensive for most individual small farmers, yet it is too crude and usually not powerful enough to serve community purposes.

Solar water heaters also exhibit low-, medium-, and high-technology levels. An example of the low-technology solar water heater is the so-called pillow type which was widely used in Japan in the 1960s. It consists of a plastic bag, with a black bottom and a clear top, which is filled with water and placed on the roof in the morning; by the afternoon the water is hot. The cost of this device was extremely low. The medium-technology solution is the flat-plate collector. The water is heated through a glass-covered panel and circulates (thermosiphoning) to a reservoir located immediately above the collector without the use of a pump. This type of solar heater is usually made of materials such as galvanized metal and has a low efficiency and short operating life. High-technology solar heaters are presently being manufactured in the United States by companies such as Grumman. These devices use high-performance and high-durability materials such as copper, are extremely efficient, and cost over $1,000. Though all three options fulfill the same task—they heat water with the sun—they are characterized by enormous differences in cost, durability, reliability, convenience, resource use, and manufacturing complexity. The high-technology solar water heater, or even the medium-technology solution, may be as inaccessible to an Asian urban squatter as a quadrophonic stereo system. For these people it is not the entire family of solar water heaters which is an "appropriate" technology, but only a particular number of low-technology solutions.

## SOLAR HEATING

The use of solar energy for heating houses has always been considered an eminently appropriate technology: it requires little mechanical energy, is by its very nature decentralized, and minimizes the use of nonrenewable resources such as coal or oil. However, on closer inspection, it becomes apparent that examples of solar house-heating technologies likewise fall into the low, medium, and high categories. All solar heating technologies are appropriate, but some are more appropriate than others.

The energy crisis prompted a renewed interest in solar house heating. Though solar houses had been built since the 1930s (one of the first was at MIT), they had never been considered as "commercial" and were the focus of attention of only a small group of researchers and inventors and, later, of the youth culture. With the prospect of ever-increasing oil prices, major industrial corporations began to develop solar hardware. Companies such as Honeywell, Inc., General Electric Company, and Hitachi Zosen International, S.A., were able to combine engineering experience and resources with access to research funds, both private and public, and soon became the leaders in the field. But the resultant technology was neither particularly simple nor particularly inexpensive. High-technology solar heating involves automation, high-performance materials, and sophisticated processes, all aimed at overcoming the diffuse, periodic, and uneven quality of solar energy. Paradoxically, solar house heating of this sort is in danger of becoming an energy-conserving technology only for the relatively wealthy.

A medium-level technology for solar house heating does exist—passive solar heating. Passive solar heating relies on trapping solar energy via large amounts of glazing and storing it in the structure itself (roof, walls, floor), which is usually of concrete, stone, adobe, or some other building material with a high thermal-storage capacity. There are various degrees of passivity, but it is generally agreed that little or no machinery or mechanical devices should be used. The passive approach has many advantages, among which is simplicity of operation, an absence of ducts, solar collectors, and heat stores, and consequently rather lower cost than the high technology—often as low as one quarter the cost of a conventional active system. Of course, there are certain disadvantages. The system is not automatic and only partially controllable, large quantities of masonry are required to store the heat, and a certain amount of temperature fluctuation is required within the house for the system to function. Nevertheless, the fact that many passive houses have been built attests to the attraction of a technology that for one quarter the cost can still provide as much as one half the required heat.

There is also a low technology for solar house heating, which could be called "solar tempering." This involves placing windows on the south side of the house as much as possible, in order to maximize the heat gain on sunny days. Some heat-storing capacity should be provided, perhaps a concrete floor, and some way of reducing nighttime loss through the windows, perhaps with shutters or heavy curtains. Solar tempering might also be achieved by attaching a greenhouse to the south wall and using this as an extension of the living room on warm winter days.

Many of these techniques involve almost no extra cost, though of course the reduction in heating cost is also less than for high-technology methods—under 20 per cent.

The existence of low-, medium-, and high-technology solutions to solar house heating points out the danger of oversimplifying the distinction between appropriate and inappropriate technologies. It is likely that passive solar heating is the most appropriate solution for many prospective home builders; it is probable that solar tempering will be the only technique that the vast majority of people can afford.

## MAIZE GRINDING IN KENYA

Bio-gas plants, windmills, and composting toilets are all machines that are designed to reduce man's labor, but there is another category of machines—those that make things. Though these two categories are often treated indiscriminately, they require separate consideration, for, in addition to labor-intensiveness, amount of capital investment, and resource use, the quality and type of product become important considerations. Thus an "appropriate" technology that produces a product that nobody wants to buy—as in the case of village soap industries in India—is seriously compromised. The existence of a demand for the quality and type of product produced must, in some way, also be a measure of appropriateness. This is obviously critical in free-market economies, but probably of equal importance to centralized economies, as the existence of black markets in many countries indicates. The emphasis on technology has sometimes clouded the fact that technologies, particularly manufacturing technologies, exist and must operate in specific economic environments. It is the over-all context of the environment which will determine what is "appropriate," sometimes with surprising results.

Frances Stewart, a British economist, has made a study of the factors affecting technological choice with regard to maize grinding in Kenya.[17] The traditional (low) technique involves using hand-operated mills. Although this is undoubtedly the most labor-intensive (and least capital-intensive) technique, it is not very productive and so physically debilitating that it is used only for home production. The two main processes for commercial maize grinding turn out to be archetypes of high and medium technology. The hammer mill (medium) is a small, locally manufactured, diesel-operated machine that eliminates the hard labor of the hand mill but is still a fairly labor-intensive technique. The roller mill (high), on the other hand, is an imported, larger-scale machine that employs one fifth the number of workers per investment unit compared to the hammer mill and costs twelve times more per unit of output.

This seems to be a classic case of the medium technology which employs more people, increases national self-reliance, requires lower investment both per employee and per unit of output, and is also finally a small, decentralized technology. What is more, it appears that in Kenya the majority of grinding is in fact done by people using hammer mills. But in her study, Stewart found that the number of hammer mills is diminishing, while the demand for the (more expensive) product of the roller mills is increasing. Why is this happening?

It turns out that, as sometimes happens, the products of two different technologies are not exactly comparable. The medium technology produces a rough-ground, un-

17 *Technology and Underdevelopment* (London, 1977).

sifted flour called *posho,* which, though extremely nutritious, does not keep for more than two or three days without spoiling. The high technology produces a sifted flour which is less nutritious, but which is better packaged and can be kept longer. The taste is also different. It appears that the consumers who possess the bulk of the purchasing power prefer the sifted flour. "The preference is partially (but by no means wholly) a reaction to advertising," according to Stewart. "[The] switch also arises from the urbanization process, since the roller mill products are more suited to an urban society, being better packed and keeping longer." The fact that people were prepared to pay more for the sifted than for the unsifted *posho* significantly altered the economic comparison between the two technologies: taking into account the monetary value of the products, the labor productivity of the sophisticated mill became *higher* than that of the hammer mill, while its investment productivity was only very slightly lower.

What conclusions can be drawn from this? Critics will undoubtedly point at the nefarious and one-sided role of advertising in shaping consumer preferences; their indignation will be further fueled by the fact that the high-technology products in this case have less protein and fewer important minerals. (It is odd, though hardly mitigating, to note that traditional hand grinding removes a similar portion of the germ and bran.) But the fact that Kenyans are beginning to prefer cleaner, longer-lasting, sifted flour is easier to criticize than to change. Germans like dark pumpernickel, the French prefer crusty white *baguettes,* Poles eat rye bread, while most Americans insist on consuming a cottony white bread. There is no accounting for tastes, or rather, manufacturing technology *must* account for them. The roles of advertising, status

seeking, images of convenience, and gastronomic evolution are complex indeed—they are both resilient and im-
mutable. The Kenyan example shows that "one cannot
draw any conclusions from the nature of production
methods involved without also looking at the implications
for consumption pattern" (Stewart). Just choosing the
"right" hardware is obviously not enough.

Whether one is a capitalist or a socialist, choosing a
technology involves making an investment decision, weighing the costs and the benefits. The most useful contribution of AT may be to broaden the definition of what
these costs and benefits should include and to point out
that employment creation should be considered a benefit
in itself and that various social costs should be added to
the equation. At the same time, the AT movement has
been almost totally involved with the cost side of the investment equation and has sometimes neglected the benefit side. Thus the high cost of bio-gas plants has been
rationalized on the vague premise that fuel and fertilizer
are "valuable." More attention paid by technologists to
the actual benefit of fuel and fertilizer to landless peasants
would have given some hint that it was unrealistic to expect them to take advantage of this technology.


## EARTH BUILDING

The use of earth for rural construction has been a traditional practice in many countries and has recently been
revived as an "appropriate technology" following the
efforts of people such as the Egyptian architect Hassan
Fathy, who documented his work in *Architecture for the
Poor* (Chicago, 1973). In many countries of the world,
under many different names (adobe in the Americas,
*banco* in Africa, or just mud), earth is used to make

bricks or monolithic walls or is plastered on frames. Its durability as a building material is a function of the type of soil that is available and of the climate. Where clay soils are available and where the climate is dry, such as in Fathy's Egypt, excellent bricks can be fabricated of earth.

It is argued that earth construction is advantageous because it makes use of a local material and the production process is labor-intensive, and hence it is considerably cheaper than other building materials, even with the addition of a small amount of cement. However, a consideration of over-all benefits does not support the view that earth construction will *always* be cheaper.

Two United Nations experts made a study of earth building in Trinidad.[18] As expected, soil/cement blocks were about 50 per cent cheaper than conventional concrete blocks. However, when the cost of a complete house was calculated, the use of soil/cement became *more* expensive than concrete blocks. Soil/cement blocks are considerably heavier than the hollow concrete blocks and, being more porous, also require more mortar. It took almost twice as long to build a house using the soil/cement blocks, thus offsetting the advantages of lower material cost. The final house in soil/cement blocks was about 50 per cent more expensive than the concrete-block house. In addition, in the climate of Trinidad, while concrete blocks could be expected to have a life of fifty to one hundred years, soil/cement blocks showed signs of deterioration after only two. Clearly, in the context of Trinidad, the benefits of the concrete block outweighed its initial higher cost. My point, once again, is not that soil/cement

[18] "Interim Report on Some Aspects of Low Cost Houses Built in Trinidad and Tobago," unpublished report of the United Nations Technical Assistance Program to the government of Trinidad and Tobago, by Alvaro Ortega and P. Selvanayagam, October, 1966.

is not useful, but only that "usefulness" must be measured by a consideration of over-all benefits, not by some narrow measure of "appropriateness." There are many situations where earth is an extremely beneficial building material, particularly when it is stabilized with a small quantity of cement and especially in rural areas where conventional cement blocks are not available and where the use of adobe or earth bricks is already a well-established building technique.

In this context it is worth mentioning the Cinva-Ram, a manually operated press for block making developed in 1956 by a Chilean engineer, Raúl Ramírez. The press has turned out to be one of the most adaptable and successful intermediate machines: it has been used throughout Latin America and increasingly in Africa. Part of the success of this device is attributable to the ease with which it can be replicated in most situations. Typically, one machine is brought from Colombia, where it is manufactured under patent of the Rockefeller Foundation, and serves as a model for subsequent local production. Cinva-Rams have been produced in local workshops in Guatemala, Tanzania, Mozambique, and Mexico and have been extensively used in self-help housing programs, where the longer erection time (of the heavier blocks) is not a major disadvantage.

## CHOOSING TECHNOLOGIES: INDONESIA AND EAST PAKISTAN

The tendency to emphasize the technique or the machine has given rise to a misconception that the bottleneck in the use of intermediate technology is always a paucity of appropriate or intermediate options. Schumacher particularly stressed this point; one of the seminal chapters in

*Small Is Beautiful* is entitled "Social and Economic Problems Calling for the *Development* [my emphasis] of Intermediate Technology." People are not using intermediate technologies because there are not enough technologies available (the argument goes); what is needed is to develop such technologies and then people will use them. I believe that there is some evidence that, in fact, intermediate technologies do already exist in a number of fields. The real bottleneck is often not a lack of choice but rather the way that the choice is made. It is crucial to understand this difference, for it may well be that promoting the use of intermediate technologies does not always involve inventing new technologies, but rather convincing people to change the way that they decide which technologies to use.

An American professor of business administration, Louis T. Wells, has studied in detail how entrepreneurs choose technologies in a selected number of industries in Indonesia.[19] These industries, which manufactured, among other things, cigarettes, flashlight batteries, soft drinks, and tires were picked because each used a range of technologies which were classified as capital-intensive, labor-intensive, and intermediate. Thus, in cigarette manufacturing, the capital-intensive technology used machines for all steps of the process; the labor-intensive technology used no machines at all, the cigarettes being prepared and rolled by hand; and the intermediate technology used a combination of hand preparation and machine rolling.

The reasons for the existence of different technologies is usually attributed to varying factor costs—that is, the

[19] "Economic Man and Engineering Man: Choice of Technology in a Low-Wage Country," in C. P. Timmer et al., *The Choice of Technology in Developing Countries: Some Cautionary Tales* (Cambridge, Mass., 1975).

costs of labor, capital, and raw materials. It is generally assumed that the entrepreneur, acting as an "economic man," will choose the technology that minimizes these costs. In Indonesia, a local industry which had trouble raising capital tended to use more labor than did foreign or state-owned industries, which had easier access to capital and hence could be more capital-intensive. Wells found that usually the existence of an intermediate technology could be justified on the grounds that additional investment per worker (for some machinery) would be paid for by higher returns. However, the existence of capital-intensive industries could *not* be explained as a result of minimizing factor costs. The additional investment per worker for automatic machinery often far exceeded any possible wage savings. Yet many entrepreneurs expressed either the intention or the desire to replace labor-intensive or intermediate equipment with sophisticated machinery. Why this apparent "uneconomic" desire to switch?

It is sometimes claimed that the products of capital-intensive technologies are superior, as in the case of maize grinding in Kenya. However, in the industries studied in Indonesia, though the labor-intensive techniques did tend to result in an inferior product, there was no appreciable difference in the quality of the products turned out by the intermediate and capital-intensive processes; indeed, in some cases the intermediate and automatic technologies were used side by side in the same plant. Neither was ignorance of the intermediate technique the reason for switching to capital-intensive processes; some plants already had intermediate technologies which they were in the process of converting. Finally, the intermediate technologies did not appear to use more raw materials than the automated production processes; in some cases they were actually more efficient.

Wells identified a number of noneconomic factors that seemed to encourage the move to capital-intensive technologies. Capital-intensive techniques could be more readily adjusted to meet different levels of demand—a machine could be slowed down but a worker had to be laid off (though Indonesian labor laws discourage the latter practice). A capital-intensive plant could also be an insurance against future price competition, since the capital-intensive process tended to have a lower marginal cost than the labor-intensive one. However, the main factors seemed to be not economic but engineering in nature. The managers reduced operational problems to those of managing machines rather than people; they aimed at producing the highest quality product possible (not just the highest quality desired by the consumer); "engineering aesthetics" played a role in choosing automated over intermediate equipment. It was Wells's conclusion that the entrepreneur acts not only as "economic man" but also as "engineering man."

There are other indications that the stumbling block to the use of intermediate technology is institutional and not primarily technological. John Woodward Thomas, a specialist on rural development at Harvard University, analyzed the decision-making process that took place in 1960–70 during the implementation of a well-drilling program in East Pakistan (now Bangladesh).[20] This huge (over 10,000 wells) program involved the East Pakistan government, the World Bank, the British government, a Swedish aid agency, a Yugoslav drilling firm, a British

[20] "The Choice of Technology for Irrigation Tubewells in East Pakistan: Analysis of a Development Policy Decision," in Timmer et al., op. cit.

supplier, and other international consultants. In spite of the fact that a low-cost type of well could have been built, *all* the various installations were medium-cost. The medium-cost solution, in a classic way, relied on foreign contractors, expensive equipment, and a low labor input. According to Thomas, the economic rate of return of the medium-cost technology was lower than that of the low-cost solution. The conclusion of this study was that the main factors governing the choice were not economic, but that "Ultimately, it was the organizational requirements of the implementing agencies, including the aid donors, that determined the choice of tubewell technology for East Pakistan. In the actual decision-making, such factors as risk avoidance, appearance of modernity, established procedures, familiar techniques, and by no means least, control, outweighed development policy objectives."

Two interesting facts emerge from the Indonesian and East Pakistan studies. First, there was a range of technological options available, both for well drilling and for manufacturing, that included, among others, intermediate technologies; in *all* cases there was no apparent need to develop new technologies. What is more, the performance of the intermediate technologies was comparable to the performance of the higher-cost technologies. It appears that the high technologies were chosen not because of lower cost, but in *spite* of higher cost.

Thus, the second fact that emerges is that the higher-cost technologies were chosen not for economic reasons. In one case, the entrepreneurs expressed a desire for modern machinery, a preference for automation, and an appreciation of "quality" in production that was engineering-, not market-, oriented. In the case of East Pakistan, a preference for modern solutions was linked to institutional requirements which favored foreign contractors over locals, familiar techniques over unfamiliar ones, centralized programs over fragmented ones.

## SOFT TECH IN AMERICA

The preference for intermediate technology options has emerged, paradoxically, in the United States. This has caught many AT advocates unawares, for, like Karl Marx who ignored Russia in the conviction that a communist revolution could only happen in Western Europe, their hopes, from the beginning, were on the less developed countries. Like communism in the Soviet Union, AT in the United States has been adopted at an unlikely time, in an unlikely place, and for unlikely reasons.

I have described how, in China, intermediate technology is being used largely as a stepping stone to industrialization; in the United States it *follows* industrialization. How can this be?

Revolutionary AT in the rich industrialized countries exists either in the minds of the neo-Utopian writers or in a very few, not wholly successful, "demonstration projects." But there is also evidence of another type of application of intermediate technology, which is neither overtly antimodern nor ideologically hidebound. I would like to use the term "soft tech," recently revived by Stewart Brand, to differentiate it from Appropriate Technology in general. Three examples of soft tech in the United States are wood-burning heaters, on-site waste disposal, and owner-built housing.

In the last five years there has been, in the United States, an extraordinary renaissance in the use of wood as domestic heating fuel. Although this has undoubtedly been prompted by the energy crisis, it cannot be explained by economics alone. The environmental movement has had an influence on public awareness of fossil fuel depletion (wood is a renewable resource), but neither can this fully explain the depth of this new interest, which is characterized, above all, by a fascination with stoves themselves.

Wood was traditionally a heating fuel in North America, and in the two centuries since Benjamin Franklin, attempts have been made to improve the efficiency and effectiveness of the wood-burning stove and fireplace. Most of these developments took place in the nineteenth century, however, and the device which has recently caught the attention of the public is the airtight stove developed in pre-World War II Scandinavia. The principle of such stoves is to control air input completely and thus maximize and prolong combustion.

A common characteristic of wood stoves and fireplaces is their relative simplicity. A few manual controls—for feeding the fire, adjusting the draft, and, of course, cutting the wood—are all that are required, for unlike the oil furnace or the electric radiator, wood stoves are rarely automated. This is, I believe, the key to their popularity. There is a satisfaction gained from this rather elemental activity over and above the economic gain and the sense of environmental decency. Critics of soft tech have sometimes ridiculed the fact that the owner of a wood stove may also have a television set or a microwave oven. They have missed the point. The wood stove is not an alternative to affluence, it is a *by-product of affluence*. It provides satisfaction which electric baseboard heaters cannot give.

A second example is the concern, at the individual level, with the disposal of human wastes. This was initially also a result of the environmental movement drawing attention to the effects of improperly treated sewage and to the inappropriateness of using large amounts of water as a transport medium for human wastes.

A technology that has been associated with domestic on-site waste disposal is the composting toilet. There are many variations of this, some owner-built and some factory-produced. A common characteristic of both types is a relatively nonmechanical device using biological decomposition (without water) for the production of humus, which can be used in the garden as soil conditioner. The reuse aspect is important, for users of composting toilets have often described the importance of knowing that they are *personally* taking responsibility for recycling their wastes.[21]

Composting toilets do not simplify life; like wood stoves, they require certain manual operations and, more important, awareness. Composting toilets are not only used for economic reasons; in fact, they are often expensive and it is no coincidence that they were developed mainly in Sweden, a country with one of the highest standards of living in the world. But composting toilets do give the individual the opportunity to do something about protecting the environment, at least at his own scale. They give, once more, a sense of satisfaction.

The sense of satisfaction plays an important role in owner-building. In *Freedom to Build* (New York, 1972), William C. Grindley, an American architect, pointed out the startling fact that in 1968 fully 20 per cent of all single-family houses in the United States were owner-built, this in spite of the growth of factory-built houses and mobile homes. The strongest impetus for owner-building is certainly economic, since the owner who builds his own

[21] See, for instance, Carol Stoner, ed., *Goodbye to the Flush Toilet* (Emmaus, Pa., 1977), or Sim Van der Ryn, *The Toilet Papers* (Santa Barbara, Calif., 1978).

home can save from one half to three quarters the cost of
a contractor. But as anyone who has built his or her own
house knows, more is involved than simply money. There
is the important satisfaction gained in having a house that
suits one's own needs (usually the prerogative of the
wealthy) and from the construction process itself. This
last is attested to by the fact that by no means all owner-
builders are indigent; many finance their homes the conventional way—by a bank loan.

What encourages owner-building in America is,
strangely enough, affluence. There are large amounts of
leisure time; there are low-cost, hand-operated power
tools (years of training are not required to acquire building skills); and there are standardized building products
(plywood, plasterboard, precut lumber) that facilitate the
task. The inventiveness and sheer elation of much owner-
building, especially that of the young, attests to the fact
that house building has taken over the role of true
recreation.[22]

However, I do not believe that these and other examples—what I have referred to as "soft tech"—are har-
bingers of "a radical transformation of human identity."[23]
Nor do they represent a turning away from modern technological society. It is much more probable that they are
technological refinements, a process of filling in areas
which large, anonymous, impersonal technology does not
reach or, on the emotional level at least, does not satisfy.

Soft tech, in its most popular manifestations, is a refine-

[22] The phenomenon of *youthful* owner-builders, as described
by Lloyd Kahn, ed., in *Shelter* and *Shelter II* (Bolinas, Calif.,
1973, 1978), is unique to the United States. Examples in Europe
are rare; examples in other parts of the world virtually nonexistent.
[23] This is the theme of Theodore Roszak's *Person/Planet: The
Creative Disintegration of Industrial Society* (New York, 1978).

ment of certain aspects of life in the Modern Age. On the whole, it is not an attempt to replace modern technology; perhaps at most, it is an attempt to add on technologies which give greater personal returns. It is important to note that soft tech coexists with modern technology and usually takes advantage of technological advances. It is quite possible that soft tech is a symptom of very advanced industrialization: an attempt to "slow down" certain personal aspects of an increasingly accelerating society, but one which is made possible by the very productivity and rationalization of that society.

## DOES AT WORK?

In the beginning of this chapter I asked the question "Does AT work or doesn't it?" though the astute reader will have noticed that I did not promise to answer it. The literal question of how well some of the appropriate technologies perform is not my major concern. AT proponents can be just as mindlessly optimistic as certain engineers can be about aerospace technology or computers, and some of the wind machines, bio-gas plants, and solar devices publicized in the press do *not* work. My concern is whether AT "works" in a broader sense. Is it a viable approach to choosing technologies? Are there any indications that it could achieve what it claims? Is it getting at the root of the problem?

The answers, in light of the previous documentation, must be qualified: both yes and no. The examples I have looked at cannot, in themselves, prove or disprove the case, but they can indicate tendencies which, I believe, should be considered.

First, they indicate that appropriateness does not mean the same thing in all contexts. *It is not possible to predefine criteria of appropriateness.* These will differ as a result of the economic and cultural context and, as Carl Riskin pointed out, as a function of scale and rate of development. Just as it is a mistake to assume that, to paraphrase former Defense Secretary Charles E. Wilson's famous comment, "what is good for General Motors is good for the United States [and the world]," it is also a mistake to assume that the use of local labor and materials, or self-reliance, or simplicity, are desirable *per se.* Further, it is a mistake to postulate that the adoption of AT will not have any unintended side effects. It appears that AT is just as prone to these as are any other technologies, and it is false to claim otherwise.

Secondly, a consideration of appropriateness must take into account benefits as well as costs. *Appropriateness should not be prejudged without a consideration of overall costs and benefits.* This may seem obvious, but, as previous examples have shown, a self-righteous attitude on the part of practitioners has sometimes minimized or ignored reduced benefits, while emphasizing reduced costs.[24] The tendency to prejudge technologies has resulted in "appropriate" manufacturing processes that produce goods for which there is no demand or in technologies that are ostensibly to benefit the poor but in fact benefit the rich.

Thirdly, there are indications that noneconomic criteria

[24] An example of ignoring costs and benefits is the proposal, often made, to use pedal-powered devices to generate electricity or to run machinery. Although the bicycle is a marvelous invention, a human being, peddling hard, can put out only about 75 watts per hour. As the *CoEvolution Quarterly* (Winter, 1978–79) points out, this amounts to an hour of rather hard work for an electricity saving of about three cents! There are very few situations where such a machine makes sense—it falls into a general category which could be called "Robinson Crusoe technology."

play a large role in the choice of technologies. *It may be that the choice of inappropriate technologies is often an institutional rather than a technological problem.* It is simply not true that the reason appropriate technology is not used is because of ignorance or lack of choice. In many cases "high" technologies are chosen with a full knowledge of the existence of an intermediate option. When people are not using intermediate technologies because the latter don't exist, there is reason to be optimistic; when they are not using them because they don't want to use them, a not necessarily hopeless but infinitely more problematic situation emerges.

It is not obvious that the creation of organizations that specialize in appropriate technology is going to break down the institutional barriers that presently exist against the use of small or intermediate technologies. If people are not using these technologies because of nonrational prejudice, then the problem is one of changing attitudes, not necessarily of inventing new technologies. If AT is isolated as a special kind of technology, it is unlikely to alter the over-all process of making decisions; it will tend to be seen as a special case.

A more successful approach, which is particularly evident in soft tech, is the provision of information on intermediate technologies *directly to the individual.* The influence of publications such as *The Whole Earth Catalog* has been paramount in changing the attitudes of individuals and of institutions; the influence of organizations such as AT International or the National Center for Appropriate Technology, because explicit and narrow, has been much less important. The provision of information through widely distributed publications also plausibly supports a number of AT ideals: it permits the individual to decide what is appropriate, it supports decentralization, and, almost by definition, it ensures that the individual establishes a healthier control over his technology. There is evidence that the approach of making intermediate technologies available directly to the individual is not restricted to the United States.[25] It could also be argued that successful AT antecedents such as rural medicine in China, the Vietnamese sanitation program, or Gandhi's hand-spinning campaign, have all been primarily *information* strategies. The decentralization of technique has been the result of the much more important strategy of the decentralization of knowledge.

[25] The Mexican Ministry of Education is sponsoring a series of do-it-yourself soft-cover books which follow the format of the popular *fotonovelas* (illustrated romantic stories). Luis Lesur, an anthropologist who is preparing these guides, points out that the *fotonovela*, whose total sales in Mexico exceed 70 million copies per *month*, is the authentic medium of communication in his country. Significantly, all previous do-it-yourself publications in Mexico have been aimed at the middle-class and have simply been translations of United States material.

## Chapter 6

## MAN/MACHINE: CONCLUSIONS

"The reason why we are never able to foretell with certainty the outcome and end of any action is simply that action has no end. The process of a single deed can quite literally endure throughout time until mankind itself has come to an end."

—HANNAH ARENDT, *The Human Condition*

The purpose of history, especially contemporary history, is to learn something about the present. The purpose of this review of the Appropriate Technology movement—of its origins, its successes, and its failures—has been to cast some light on a number of important issues, among them the relationship between industrialization and development and between technology and ideology. I have tried to examine these questions on the basis of documented experiences; the results are not conclusive. They have not, indeed they could not have, proved or disproved all the extremely general claims that have sometimes been made. On the other hand, they have clarified certain aspects of the way that technology, especially small-scale technology, affects the way that people live.

I have previously differentiated between two views of technology: the evolutionary and the revolutionary. The former proposes technology as a stepping stone to modernization, the latter suggests a new kind of technology with different social and technological goals. Although there is considerable overlap between these two tendencies, I shall deal with the evolutionary and revolutionary aspects separately, but first I would like to raise three issues that are common to both positions: the ideas of social reform, self-reliance, and nonviolent technology.

## SOCIAL REFORM

I have already described a number of experiences which indicate that social reform through technological change alone is unrealizable. The corollary is also true: technological change without social reform will likely serve to exacerbate, not ameliorate, social injustices. The idea that social reform can be accomplished by "twiddling the technological knobs" is an attractive one, not the least to national and international technocrats. However, true social reform implies changing traditions, cultural habits, political institutions, and often human attitudes; it is a long, difficult task and there is no guarantee of success. It is true that technology can aid, even accelerate, the process (e.g., socialism plus electricity), but it cannot accomplish it alone. Furthermore, it is technology which must complement social reform and not vice versa. It is unlikely that social reform which is instituted only to facilitate a technological change will have much chance of success.

The distinction between technological change and social reform is crucial. It is naïve to hope that the introduction of a socially appropriate technology to a repressive regime will (surreptitiously) create a climate for social reform. It is worse than naïve to claim that the use of a particular technology will result in social justice or more equitable distribution of wealth in societies where

neither of these are social or economic realities. To insist
that the success of a technological program, say, rural
water supply, should rest on as yet unachieved social reform is to hopelessly muddle priorities.

I believe that historical experience has shown that social reform may follow technological change (for example, in nineteenth-century Britain), technological change
can occur with the minimum of social reforms (for example, in present-day Soviet Union), and that social reforms
may occur with the minimum of technological change (for
example, in Tanzania in the 1970s). Just as social reform
is not a precondition for technological change (in South
Africa, for instance), so technological change is not a
precondition for social reform. It is true to say that poverty cannot be redistributed—a poor democracy or a poor
dictatorship is still a poor country—but it is also true to
say that the effects and benefits of technological change
will be different in one than in the other. Technological
change may benefit the elite, the middle class, or the
poor, but this will be a function of the preexisting social
conditions, not a function of the technology itself.


## SELF-RELIANCE

There has been a great reemphasis recently on self-
reliance, which is held to be a great advantage at virtually
every level. The individual should be self-reliant in energy, the small town should be self-reliant in employment,
the urban neighborhood should be self-reliant in food
production, the nation-state should likewise be self-reliant
in food and energy. One might well ask: "Where does
self-reliance end? Or begin?"

*Thou shalt be self-reliant.*

Perhaps the reader feels I exaggerate, but the concept

of self-reliance has acquired religious overtones. It seems that self-reliance is always and in every way desirable; one should be as self-reliant as possible in every situation. There is an Institute for Local Self-Reliance in Washington, D.C., a *Guide to Self-Sufficiency* (published by *Popular Mechanics* magazine), as well as a newsletter for American Indians called *Native Self-Sufficiency*.

Self-reliance is, at first sight, an attractive concept. For the individual it offers a withdrawal from the bureaucratization of public life, for the neighborhood it promises separation from big-city politics, for small towns it raises bulwarks against the invasion of big-city commerce, and for the nation-state it promises greater political autonomy.

All these promises are illusory.

Self-reliance at different levels *simultaneously* is a patent impossibility. National self-reliance might preclude regional self-reliance, and it might well involve specialization: California or Alberta produce oil, Florida produces oranges, Nova Scotia produces fish. Likewise, regional self-reliance may negate neighborhood self-reliance and individual self-reliance. And so on.

Self-sufficiency has played an important role in the ethos of the youth culture. Though there were some early critics of the "destructive fantasy" of self-sufficiency, these were rare.[1] Most publications, like *The Whole Earth Catalog,* actively propagated the idea that the individual could, and should, be as self-sufficient as possible.[2] I have

[1] Peter Van Dresser, an American solar inventor since the 1930s, interviewed in 1973: "The drop-outs I criticize most are the ones that *pretend* at self-sufficiency—living in a wigwam and all the rest of it—yet going to Safeways once a month for their proteins. This is just a destructive fantasy" (Peter Harper et al., eds., *Radical Technology* [New York, 1946]).

[2] Stewart Brand has recently reversed his position on self-sufficiency: "Self-sufficiency is an idea which has done more harm

already pointed out the attraction of self-reliance to the do-it-yourself consumer; much of this attraction was grounded in the age-old American ideal of Rugged Individualism. Most of the youth culture's attempts at self-reliance—the communes, the free schools, the alternative industries—were short-lived and, to the extent that they were disassociated from their social environment, unsuccessful.

There is a case to be made for discriminating decentralization, as Amory Lovins had done in *Soft Energy Paths*, arguing that solar house heating has many technological and economical advantages over network systems. However, I do not think that this validates the *over-all* approach to self-reliance. Solar house heating is a special case that must be judged on its own merits. It is also largely a function of the fact that the most logical way to utilize an energy source that is by nature diffused is in a dispersed manner.

The argument for self-reliance is even harder to support on the international level. One of the important international effects of the Industrial Revolution was the development of steam power, steamships, and relatively low-cost, long-distance transport (sailing ships, though they used wind power, required very large crews and long voyages and were consequently very expensive). As a result, a whole range of one-time luxury goods such as coffee, tea, and bananas became international mass con-

---

than good. On close conceptual examination it is flawed at the root. More importantly, it works badly in practice. Anyone who has actually tried to live in total self-sufficiency—there must be now thousands in the recent wave that we (culpa!) helped inspire—knows the mind-numbing labor and loneliness and frustration and real marginless hazard that goes with the attempt. It is a kind of hysteria." (J. Baldwin and Stewart Brand, eds., *Soft-Tech* [Sausalito, Calif., 1978]).

sumer goods. This had, and continues to have, an important effect on many of the tropical countries. The mainstay of many of the less developed areas is still this kind of agricultural export: rubber (Malaysia), cocoa (West Africa), tea (Kenya, Sri Lanka, India), coffee (Brazil), cotton (East Africa), sugar (Cuba), and so on. In all these cases, the expansion of the export industry dates from the turn of the century or later. The role of trade in development is obviously important in the case of the previously stagnating sheikhdoms of the Middle East, as well as in other oil-rich countries such as Nigeria, Iran, Indonesia, and Venezuela. Export trade of manufactured goods has been the foundation of the development of Japan, Taiwan, and South Korea, and, increasingly, of India and Brazil.

Thus we see that it is the very *lack* of self-reliance on the part of various advanced countries that has permitted the "latecomers" to develop as quickly as they have. In fact, many observers (including most of the less developed countries) feel that one of the major obstacles to more rapid development is the high tariffs and trade restrictions that most of the industrialized countries have maintained on manufactured goods. It can thus be seen that national self-reliance on the part of an industrialized country (by manufacturing synthetic rubber and cotton and substitutes for imported foods) will have an extremely negative effect on many of the countries which are in the process of developing.

Self-reliance, whether for the individual or the nation-state, is, finally, a chimera. It is impossible to achieve, except in the most primitive of worlds. Buckminster Fuller said, *"All* of humanity is dependent on *all* resources." Even if the world moves to renewable energy sources, this will still be true: metal alloys for wind-machine blades, copper for flat-plate collectors, potash for glass covers. Interdependency is the unavoidable fact of the Modern Age.


## NONVIOLENCE

Most writing on technology that emanates from the international organizations such as the World Bank, the United Nations agencies, and the OECD studiously ignores the discussion of nonviolent technology altogether; one senses an embarrassed reluctance to raise the topic. This is also true, to a lesser extent, of various national groups which delicately avoid using this contentious term. E. F. Schumacher, on the other hand, used the word consistently and forthrightly, and I believe that it is an important concept, which it would be a mistake to ignore.

Within this context, nonviolent technology implies a number of things—above all, a technology which will not have unintended side effects; that will not cause social disruption; and that can be insinuated alongside traditional techniques without disturbing them. It implies a technology that is *completely* under human control. This is the social face of nonviolent technology. There is also a more ephemeral physical face that has to do with technology that will be nonviolent with respect to nature; that will not destroy, exploit, or manipulate the natural world. It is claimed that such a nonviolent technology is possible and that, indeed, it is a crucial criterion for appropriateness.

The experience of many nontechnological societies has been that the introduction of technology is a socially disruptive process. Change, sometimes violent change, inevitably follows technological innovation. Cultural habits and living patterns which have lasted for centuries can be altered in decades. What is more, though some of the effects are intended, there are inevitably *un*intended effects. It is sometimes claimed that intermediate technology could be introduced into technologically backward societies with no negative consequences and no unintended side effects. I believe that it is difficult to show that *any* technology introduced into a backward (or advanced, for that matter) society is not going to have unintended consequences, and, in the case of previously nontechnological societies, these consequences are likely to have major social repercussions.[3]

To illustrate, rather than to prove, this point, I would like to cite one example which concerns the Skolt Lapps of northeastern Finland. The situation of the Lapps resembles somewhat that of the native people of northern Canada, but unlike the latter whose main occupations are hunting and trapping, the main economic and cultural focus of the Lapps' life has been reindeer husbandry. These animals, which are periodically herded together, provide food, clothing, and, when tamed, transportation. In the last two decades, however, the relationship of the Lapps to their herds has been drastically altered by the appearance of a single device—the snowmobile.

Pertii J. Pelto, a Finnish anthropologist, has studied the effects of the snowmobile on the economic and social life of the Skolt Lapps and has concluded that this technology, which was freely chosen for a specific reason, has had a number of significant, unintended side effects.[4] The

[3] At the time of writing, a number of countries in the Middle East, particularly Iran, are undergoing social upheavals which are due, in part at least, to the rapid and uneven introduction of advanced technology and advanced social concepts (sexual equality, religious equality, land redistribution) into extremely traditional societies.

[4] *The Snowmobile Revolution: Technology and Social Change in the Arctic* (Menlo Park, Calif., 1973).

most striking effect is the changed relationship between
the herdsman and his animals; traditional reindeer herding required a long time and depended on success in taming, to a certain extent, the wild reindeer. Snowmobile
herding is a rapid activity that does not require, indeed it
does not allow, this kind of man–reindeer relationship.
A parallel exists to the contemporary western cowboy of
the United States and Canada who, from his Jeep or heli-
copter, presumably no longer sings to his cattle. In the
Skolt case, Pelto claims that the mechanized herding,
which resembles stampeding more than herding, may also
have affected the health and size of the herds.

Human interrelationships have also been affected by
the snowmobile. Social contacts in northern Scandinavia
have been intensified by faster and easier travel. Social
structure within the Skolt community has been changed;
whereas traditional reindeer herding favored the older
men who had acquired the necessary woodcraft, mechanized herding gives equal advantage to the younger men.
The cash costs of snowmobile maintenance have introduced new economic constraints, and Pelto has also observed a social stratification which previously did not exist
between the Lapp families who have successfully adapted
to the new situation and those who have not.

My point is not whether this technology has improved
or deteriorated the life of the Lapp people. Pelto himself
is ambivalent about this. He feels that something has been
lost in the Lapp culture, but he also admits that "obvious
marks of deterioration, such as serious alcoholism, family
disorganization, and violent crimes, have not appeared."
He concentrates on the snowmobile as a vehicle for
change, but does indicate that almost all of the Lapps also
own chain saws, telephones, and outboard-motor boats
for lake travel in the short summers, and a good many

use electricity for cooking and lighting and have washing machines. The important point is that what appeared to be a small technological change in the 1950s has had many side effects, few of which were intended and some of which will probably irreversibly change the nature of Lapp society.[5]

Perhaps all would not agree that the snowmobile is an "appropriate" technology, though it does exhibit many of the characteristics: it is small, easy to operate and maintain, encourages decentralization, and is not very expensive. But in any case, this example does illustrate the violent effects that a new technology, even on a relatively small scale, can have on the lives of a "traditional" community. That this technology was not imposed but freely chosen only compounds the argument.

This is but a single example; it would be possible to give others. The invention of inexpensive techniques for producing nails in the early 1800s facilitated the fast and cheap method of house building known as "balloon framing," which in turn encouraged the rapid settlement of the American West, often a violent process. The introduction of mechanized well drilling in the desert countries of the West African Sahel has had a violent effect on the migrations and on the very culture of the nomadic Tuareg. It is the rare case indeed where a technology has not had some

[5] The Lapps, like native peoples in the United States and Canada, have been in a process of change for the last hundred years. In the case of the Skolt Lapps, this includes displacement, in 1940, from their traditional home in what had been Russia to Finland and the establishment, with government help, of new settlements, as well as of new reindeer herds. The Lapps enjoy all the benefits of the Finnish welfare state, including free medical services, schooling, old-age assistance, and unemployment insurance. In this context, the changes induced by the snowmobile are part of a process, not an isolated event.

violent results, and I would suggest, as Hannah Arendt
has done, that *all technology is violent.*

This is a contentious point, but then, the concept of
nonviolence is primarily philosophical. To make something, whether with tool or machine, is to wreak violence
on nature: to destroy a tree and cut it into boards, to
wrench stone out of the ground and carve it into blocks,
to melt ore into iron. The activities of technology, from
the very beginning, have been violent ones, and with the
first architect and the first engineer the struggle for mastery over gravity, over natural forces, over nature itself
begins. There is violence in the domestication of animals
no less than in the exploitation of natural forces. Philosophically, at least, the concept of nonviolent technology
seems to be a contradiction in terms.

It might be useful to point out at this point, however,
that there *is* a distinction between machines and tools.
There is a certain amount of overlap between very simple
machines and very complicated tools, but in most cases
the difference is readily apparent. Whereas the tool is precisely an "extension of the hand," the simplest machine
already contains some measure of automation. It seems
likely that the impact of technological innovation on a
backward society will be reduced when it takes the form
of tools, even sophisticated ones, rather than machines,
even very simple ones. It may turn out to be more useful
to differentiate between tools and machines rather than between violent and nonviolent, or small and large technologies. The perils to a rural society of introducing a windmill or a water-driven turbine are probably just as great as
those of introducing a diesel generator. The fact that a
windmill does not use fuel may be less significant than the
fact that it, like the diesel, is a machine, not a tool.

It is necessary to recognize that a nonviolent technology may be an impossibility. This will be difficult, for so much of the attraction of AT lies in its appeal to virtue and in the promise that here, at last, is a technology that will not have *any* unintended effects and will not push and mold traditional societies in a violent way. Practitioners of AT must be just as vigilant as regards its unintended consequences as with any other technology. There is no immunity against technological fallout with any technologies. But practitioners should not be discouraged if they discover that "appropriate" technologies may have just as violent impacts as any other kind of technology. AT should not induce any complacency—the violence should be controlled or at least minimized as much as possible, but this can only be done if it is expected. As it pulls the trigger, AT should not be surprised by the bang.

## REVOLUTIONARY TECH

By now it should be clear to the reader that Appropriate Technology is a proposition rather than a *fait accompli*. If this is the case, what general conclusions can be drawn about the future possibilities of AT? To answer this question it is important for us to make the distinction, once again, between the evolutionary and the revolutionary tendencies. Whereas evolutionary tech will be a force for positive change, revolutionary tech is on the whole, I believe, a red herring.

Various critics have attempted to exorcise modern technology by inventing labels for a "new" technology: Convivial Technology, Utopian Technology, Alternative Technology—they have had considerably less success in actually inventing the technology. This should not be surprising. Langdon Winner pointed out in *Autonomous Technology,* in 1977, that "even if one seriously wanted to construct a different kind of technology appropriate to a different kind of life, one would be at a loss to know how to proceed." The technology that is in use today is the result of scientific, philosophical, and cultural history. What is the "new" technology to be based on? Astrology, superstition, magic? How can it fail to rely on existing institutions and ways of thinking, and how, as a result, can it claim to be capable of developing in a different direction?

The best that can be suggested often consists in going back to a sort of preindustrial Arcadia. But how can one go back to a medieval decentralized feudal society and not go back to serfdom, warlocks, and the Divine Right of Kings? And in any case, did not the Middle Ages finally lead to the Industrial Revolution? Ivan Illich suggests a kind of restrained Luddism, but does not say who will control it. Who will be this Big Bully who will keep rein on technological development?

The biggest obstacle to the development of a different type of technology is the ubiquitous presence of a modern technology which is shaping human consciousness and behavior, structuring society, and determining the choices that are available. *The Whole Earth Catalog* and the other publications of the youth culture would have been impossible without the technology of Polaroid cameras, Xerox copiers, and International Business Machines computers. The success of *Small Is Beautiful* was partly due to inexpensive printing techniques and a worldwide distribution system. The fact that so much of the ethos of the "counter-culture" has been adopted by middle-class America, and not the least by many American corporations, is not so much an indication of the failure of the flower children as of the fact that they were part and parcel of modern American culture, *not* any sort of an alternative. The ease with which modern industry has been able to co-opt many of the soft technologies is an indication that modern technology doesn't just occupy the high ground—it occupies *all* the ground.

Although some have claimed to have developed a "new technology," all that they have been able to do is add to existing technology, which, while not bad in itself, is hardly the same thing.

One can only conclude that the invention of a "new" or "alternative" technology is, philosophically and practically, an impossibility. It is useful to criticize modern technology, but it is misleading to imply that another, totally different approach to technology is possible. Neither is dismantling present-day technology—surely a precondition to any new approach—a serious possibility. The wide-ranging and sometimes devastating effects of a brief power blackout or even a short-lived drop in fuel supplies are only hints of what such technological demolition would entail.

## EVOLUTIONARY TECH

The American critic H. L. Mencken once wrote that although it was unquestionably noble to die for an idea, how much more noble it would be to die for an idea that was true. A great part of the writing on the subject of Appropriate Technology has been more concerned with the former than with the latter. I have tried to show why this has been the case.

A successful approach to the development and application of small technologies should, as I have pointed out, reassess the desirability of self-reliance and seriously question the concept of a nonviolent technology. Likewise, it must be aware of the limits of technology—what it can and cannot accomplish.

There are no simple remedies; small is not always beautiful,[6] local is not always better, and labor-intensiveness is not always desirable. Small technologies cannot avoid traditional economic strictures; a manufacturing process must produce useful products; an investment, no matter how small, must bring some increased benefit.

It is sometimes claimed that the concept of Appropriate Technology contains nothing new and that "it rests on fundamental principles of benefit-cost analysis."[7] This statement is partly true. AT has not developed any new methods for determining what is and what is not appropriate. Several writers have attempted to develop lists of "criteria" for appropriate technology, but on closer inspection these turn out either to be too vague ("cultural adaptability," "ecological responsiveness") or too general ("small," "labor-intensive") to be of much use except as slogans, which, of course, is what they are. The limitation of these criteria becomes evident when one attempts to apply them to a particular situation: either all technologies are appropriate for something, in which case there

[6] It is likely that I will be accused of quibbling. Schumacher himself had little time for "those who get stuck on words; who start arguing with me that small is not *always* beautiful . . . these people who can't get beyond the words. This I consider an academic disease which is rampant" (National Film Board of Canada interview, 1977). This statement is rather ingenuous; since it is words that have fueled the Appropriate Technology movement, a critique of AT cannot avoid, in part at least, focusing on "the words."

[7] P. Rosenfield et al., "The Appropriate Technology Bandwagon: Transfer of Knowledge and Community Water Supply," paper presented at the Second International Conference on Transfer of Water Resources Knowledge, Fort Collins, Colorado, June 1977.

is no such thing as AT, or not all appropriate technologies are applicable all the time, in which case are they really "appropriate"?[8]

On the other hand, in the context of the technological development of less developed countries, particularly of the poorest ones, AT has drawn attention to the fact that many more technical options exist than was previously thought. The options that the AT movement has helped to bring to light are either those which have been prematurely discarded, or less well-known techniques which are still in use in various developing regions, or innovative technologies that have been developed by relatively obscure groups and individuals. AT has brought together a spectrum of "poor man's tools" and it has begun to break down some of the institutional prejudices against intermediate or small-scale techniques.

Perhaps the most important role that the AT movement has played in international development has not been as the inventor of a new approach, but rather as a reminder to the international development establishment that a very large number of people have been left out of the development process and that technological options do exist which could begin to rectify this situation. However, as an attempt to demodernize technology and to take an alternative path, Appropriate Technology is doomed to failure. It is a pretentious, romantic, even poignant attempt to stop the ocean with a child's beach shovel and play bucket.

So strident have been the demands to develop a new technology, and so eager has been the public to believe

[8] The choice is even more strained for most AT groups, which are usually experts in some particular technology (solar, wind, bio-gas), and for whom the process of identifying the "appropriate" solution must be a foregone conclusion.

that the solution to the perceived ills of the Modern Age lies in changing horses in midstream, that these paper heroes have been believed. And now what? What if workers don't like laborious machines? What if bio-gas plants benefit the rich and not the poor? What if wind machines are often too expensive? What if the solar heater falls apart after six months? What if no one wants to buy homemade soap? What if . . . AT cannot deliver the goods?

# EPILOGUE

In the decade since this book was first published, the appropriate technology bandwagon has traveled a rocky road and lost many adherents. First to go were the alternative energy proponents. During the 1980s, the falling price of petroleum resulted in a loss of interest in alternative energy souorces such as solar and wind power and a virtual cessation of government support for research in those fields. Simpler energy conservation measures, such as smaller, more efficient cars and higher building insulation standards—not autonomous houses—were the chief legacy of the energy crisis. In areas where electricity was expensive, solar water heaters and energy-conserving shower heads became common, as did heat pumps—no significant social changes accompanied these technical innovations. The effective moratorium on the construction of new nuclear power plants in the United States (although not in Europe) also removed the chief bogeyman of the soft energy advocates.

Nor did political developments in the 1980s help AT. First there was the election of conservative governments in the United States, Great Britain, Canada, and West Germany. Although conservatives stressed private entrepreneurship, the social climate was uncongenial to Buddhist economics—people wanted more consumer goods, not fewer. As if that were not enough, the left-of-center reformist tendencies of AT were dealt a crippling blow by events in the Soviet Union and Eastern Europe, where consumer capitalism was clearly in the ascendancy.

Appropriate technology as a tool for international development fared as badly, partly because self-reliance proved to be an unworkable chimera and partly because, in those parts of the Third World where AT had been most strongly promoted, such as Africa, economic development in the eighties was a disaster. The national and international aid agencies, notoriously fickle in their search for new technological buttons to push, have largely abandoned AT. The inability of intermediate technology to promote real economic growth was underlined by the success of conventional technology in countries such as India, which achieved surpluses in food production thanks to the Green Revolution, and in the economic powerhouses of southern Asia: Singapore, South Korea, Taiwan, and Hong Kong. The Chinese flirtation with small, decentralized technologies (described in chapter 3) has been exposed as even more disastrous than previously imagined. I had the opportunity to visit China in 1986, which confirmed my suspicion that the present modernization goals of that country differ little from those of its neighbors.

What I called the youth wing of the AT movement has grown up, and its preoccupation with radical technology can now be viewed for what it was: a fad. Although a few stalwarts continue to soldier on—Steve Baer now runs a company that manufactures a variety of energy-conserving devices—the chief interest today is distinctly on high rather than low technologies, on the personal computer rather than on the composting toilet and the geodesic dome. In 1984, the creators of the *Whole Earth Catalog* published the *Whole Earth Software Catalog,* and recently Stewart Brand organized a series of workshops on virtual reality—computer-simulated environments.

Should AT be relegated to the dustbin of history, along with other utopian ideas such as turn-of-the-century anarchism, Esperanto, and Reaganomics? I think it is too early for that. In the last several years there has been a growing interest in something called "sustainable development," which bears a marked resemblance to the environmental concerns voiced by E. F. Schumacher in *Small Is Beautiful*. Indeed, in many ways SD, while more broadly based, is a made-over version of AT, but spurred on less by economic and social issues than by news of environmental degradation, chiefly, or at least most dramatically, by reports of global warming and resource depletion. The latter problems are real enough, but there is a danger that the proponents of SD (which now include many corporations) will fall into the same trap as their predecessors, and that catchy slogans, intended to rally public support, will take the place of thoughtful analysis. Were that to be the case, history will repeat itself, and a great deal of effort will be expended to little effect.

*W.R.*
*Key West, December 1990*

# INDEX

*Drop City* (Rabbit), 90 n.
Duvalier, François, 54 n.
Dymaxion automobile, 87
*Dymaxion World of Buckminster Fuller, The* (Marks), 89

Earth building, 135–37
East Africa, 156
East Pakistan, 137, 140–42
Eckaus, Richard S., 1 n.
Ecology, 2
*Economic Development* (Galbraith), 53–54 n.
"Economic Man and Engineering Man: Choice of Technology in a Low-Wage Country" (Wells), 138 n.
Egypt, 11, 62
Électricité de France, 85
*Electric Kool-Aid Acid Test, The* (Wolfe), 91
Elizabeth II, of Great Britain, 29
Ellul, Jacques, 14, 34
El Salvador, GNP, 42
*Encyclopédie* (Diderot), 92, 94, 95
Energy, aeolian (wind), 85, 103
  bio-mass, 85
  nuclear, 85
  renewable resources, 7–8, 34, 85, 143, 156
  soft technologies, 34–35, 85, 142–46
  solar, 85, 103
  tidal, 85
Energy crisis, 97–98, 100
Environment/environmentalists, 2, 20, 21, 25, 34, 36, 57–58, 85
*Environmentally Appropriate Technology* (McCallum), 1 n.
Ephron, Nora, 98

Ethiopia, wind machines, 128
Evolutionary technology, 83, 84, 151–52, 164
Expo 67 (Montreal), 104

Fan mill, 128
Fathy, Hassan, 135
Fawcett, Farrah, 60
*Feminine Mystique, The* (Friedan), 13
Filippini, Armando, 128
Finland, 158
Firestone-India, 122
*Food First* (Lappé et al.), 56 n.
Ford, Henry, 19, 20
Ford Foundation, 10
Ford Model T, 19
Foreign aid, 61–66
  bilateral, 62, 114
  multilateral, 62
Founex Report, 57
Four Modernizations (China), 80–81
France, 23, 85
*Freedom to Build* (Grindley), 144
Friedan, Betty, 13
Friends of the Earth, Inc., 34
Fuller, Richard Buckminster, 33, 87–92, 94, 96, 99–100, 101, 102, 103, 156
  and Schumacher, 104–9
"Funk Architecture" (Voyd), 90 n.

Galbraith, John Kenneth, 53–54 n.
Gandhi, Mohandas, 36–39, 49, 117, 123
Garcia, Kjell Laugerud, 54 n.
Gardening projects, 116 *and* n.
Garg, M. K., 4, 121 n.
General Electric Company, 130
General Motors, 147
Geodesic domes, 88–92, 95, 96, 100

# ILLUSTRATION CREDITS

*Architectural Design:* 4
Vikram Bhatt: 28
Brace Research Institute: 16, 23, 24, 26, 31
International Development Research Centre: 18, 20, 21, 25
Keystone Canada: 1
Luis Lesur: 15
Alvaro Ortega: 6, 7, 17, 33, 34
Martin Pawley: 11
Witold Rybczynski: 27, 29, 30, 32
Bing Thom: 8, 9, 19
United Nations: 5
The White House: 2